PRAISE FOR RADI

"Marketing has told us that everyone should have wireless computers, wireless indoor phones, wireless mobile phones. Very early, I coined this as "the most gigantic human full-scale experiment ever". Is anyone really seriously believing this is safe? The authors and I do not. To find out more about this topic, read this book - it is written by two genuine experts but for any layman to easily enjoy. The authors explain the current science so well so even I - as a scientist - learned more!

- Dr. Olle Johansson, Ph.D., Associate Professor in Neuroscience, the Karolinska Institute, Stockholm, Sweden

"*Radiation Nation* sheds light on the vitally important issue of EMF radiation in our current world. We are just beginning to understand the far-reaching effects of what will become known as the "smoking gun" of our generation."

- Katie Wells, WellnessMama.com

"*Radiation Nation* is very useful for those who are extensively using a cell phone and other sources of EMF. I highly recommend this book for consumers as well as radiation scientists."

- Dr. Kedar N. Prasad, Ph.D., Author of over 25 books and world expert in radiation biology.

"EMF safety can be a controversial and confusing topic, but *Radiation Nation* unpacks it all with ease and clarity. You'll walk away with a solid understanding of EMF risks as well as concrete ways to protect you and your family. This book is a must-read for our digital world!"

- Genevieve Howland, Founder of MamaNatural.com

"When I first interviewed Daniel DeBaun on my podcast, I realized quickly that he is one of the most knowledgeable minds on the face of the planet when it comes to defending yourself from the host of electrical pollution that we now face in our post-industrialized era. Problem is, until I got my hands on *Radiation Nation*, I could never find one, centralized, convenient location to fill me in on all the existing research on EMF along with every single practical recommendation that exists to actually defend yourself against this radiation. But this book contains everything that you need to know, from the information to explain to your friends why you might not be holding yourself on up to your ear, to things you never would have realized about common items like laptops and smart meters, to handy, easy-to-understand tip sheets and beyond. I cannot recommend this book highly enough if you care about your own health and that of your families in our modern era."

-Ben Greenfield, BenGreenfieldFitness.com

"Finally, a handbook that explains the health impact of radiation from our technology and practical steps for protection. EMF Radiation is the new Glyphosate: it was made widely available with inadequate safety testing, we are learning too late the detrimental effects on our environment and human health. We must self-educate and self-advocate to protect ourselves from overexposure to EMFs. *Radiation Nation* is the place to start."

- Lauren Geertsen, Founder of Empowered Sustenance and Co-Founder of Meo Energetics

"*Radiation Nation* is a must read for everyone who uses mobile phones, notebooks or laptop computers. It's an absolute must read for all health care practitioners! As a nutritional therapist I have already seen the effects of EMFs on blood sugar regulation and hormones and the potential epigenetic effects on our children and grandchildren can only be described as terrifying. The good news is that *Radiation Nation* not only helps us to understand the dangers but also teaches us how to safely use the technologies we have grown to love."

- Gray L. Graham B.A., NTP, President and Founder of the Nutritional Therapy Association, Lead Author of "Pottenger's Prophecy, How Food Resets Genes For Wellness or Illness"

"Nearly all of us living in civilized society today are surrounded by WiFi and other forms of electromagnetic radiation on a 24/7 basis. It is critical to clearly understand the sources of this exposure and minimize risks where and when possible. *Radiation Nation* explains both the problem and the solutions each of us needs to know to protect our long term health from development of EMF related disease. This information is particularly important for households with growing children for whom excessive exposure to EMFs has become endemic."

- Sarah Pope, The Healthy Home Economist

"I wish everyone would read *Radiation Nation* - I believe that a great number of diseases could be mitigated by reducing our exposure to electromagnetic radiation from our devices."

- Robin Shirley, Founder, Take Back Your Health Conference

"*Radiation Nation* is a game changer, presenting the facts about radiation, not only in our homes but schools, workplaces and public places. Radiation is everywhere. Thank you for your tireless effort to educate on the facts, and for providing simple strategies to reduce radiation exposure."

- Joanette Biebesheimer, MS Biochemist, CEO No Sweat Global

THE FALLOUT OF MODERN TECHNOLOGY
RADIATION NATION

DANIEL T. DEBAUN
RYAN P. DEBAUN

This book is intended as a reference volume only, not as a medical manual. The information given here is designed to help you make informed decisions about your health. It is not intended as a substitute for any treatment that may have been prescribed by your doctor. If you suspect that you have a medical problem, we urge you to seek medical help.

The information in this book is intended to stimulate scientific inquiry and serve as a launch pad for healthy discussion. All readers are encouraged to conduct due diligence with the information presented herein, especially as it relates to matters affecting personal health. Readers are encouraged to make well-informed decisions based on all available resources as well as good judgment.

Mention of specific companies, organizations, authorities, or people in this book does not imply endorsement by the author or publisher, nor does mention of specific companies, organizations, people, or authorities imply that they endorse this book, its authors, or the publisher.

Although the authors and publisher have made every effort to ensure that information in this book is correct at press time, the authors and publisher do not assume and hereby disclaim any liability to any party for any loss, damage, or disruption caused by errors or omissions.

Internet addresses and telephone numbers given in this book were accurate at the time it went to press.

For questions or information about purchasing books, please contact:
info@radiationnationbook.com

Publisher: Icaro Publishing

Publishing services provided by: Archangel Ink

Cover Design: Ryan P. DeBaun

Interior Images: © Ryan P. DeBaun unless otherwise indicated.

Library of Congress Control Number: 2017903210

ISBN: 978-0-9981996-0-3

1. Health 2. Science & Technology 3. Education

First Edition Printed in the United States of America

www.RadiationNationBook.com

ACKNOWLEDGEMENTS

Our extreme gratitude to everyone who made this book possible including Archangel Ink, Grace Chen, family and friends.

For their kindness and contributions, we would also like to thank Dave Asprey of Bulletproof; Peter Bauman; Lauren Geertsen, Empowered Sustenance; Ben Greenfield, Ben Greenfield Fitness; Gray Graham, President and Founder of the Nutritional Therapy Association; Genevieve Howland, Mama Natural; Olle Johansson, Associate Professor in Neuroscience, the Karolinska Institute; Jeromy Johnson, EMF Analysis; Sarah Pope, The Healthy Home Economist, Robin Shirley, Founding President of Take Back Your Health; and Katie Wells, Wellness Mama.

CONTENTS

PREFACE

A few years ago, my sons were visiting my wife and me over the holidays. As many employees are doing nowadays, they were working remotely for their jobs, typing away with their laptops in their laps for hours at a time. I still remember it like yesterday when my wife said to my boys, "That can't be good for you! I want grandchildren one day!" What she said immediately caught my attention and something clicked in my mind. Now, my wife had no science or medical background, but her intuition told her that whatever "radiating stuff" that was coming from the bottom of the laptops probably wasn't going to help her sons have children any time soon. And she was right. From my engineering background at Bell Labs and AT&T, I knew that all electrical equipment and electronic devices emit "low levels" of radiation called electromagnetic radiation (EMR), also referred to as electromagnetic fields (EMFs). For the purposes of this book, it is just "radiation" or "EMF radiation."

When I worked in the engineering labs during the late 1970s and 1980s, EMF radiation was pretty low on the totem pole of things we worried about. We were more concerned about electromagnetic interference between electronics than we were about the health effects of EMF radiation on people.

Back then, though, people weren't using computers in their laps for hours at a time. To be completely honest, although I sensed there was some kind of issue with laptops and cell phones, it was not something I pondered too often. But when my wife said those words, it was like a lightbulb turned on.

So, I started to research and found that there was an extensive amount of evidence showing that radiation emissions from laptops, cell phones, and other devices had negative biological effects on humans. Actually, there was a lot more out there than I thought there would be. As an engineer, I know about materials and processes to block radiation emissions from electronics. So naturally, I figured a product had to exist that my sons could use under their laptops to shield the radiation emissions. I searched and searched online, but to my surprise, I could not find anything that completely blocked the radiation.

In fact, it didn't seem like anyone really understood what the issue was or what they were doing. You see, almost all modern electronic devices like laptops, tablets, and cell phones emit two kinds of EMF radiation: extremely low frequency (ELF) and radio frequency (RF) radiation. If you truly want to protect yourself, you must completely block both these types of radiation. From what I could find in a small set of existing products, only one type of radiation emission was blocked (either ELF or RF)—and it was not blocked very effectively, for that matter. For example, products that I found could only block about 75–85 percent of one type of radiation emission. As for the other products out there, almost everything had unsubstantiated claims or used some wacky gobbledygook explanations that had no basis in science. Here we were, way into a mobile society, and we have no effective solution for completely blocking EMF radiation.

That's when I took it upon myself to create a solution. If nothing existed, I was going to fix that. I worked on the problem for a few months, and after a lot of trial and error, I was able create a process to block all forms of EMF radiation emitted from the bottom of a laptop. I began by making shields for my sons and then for some friends and family. Everyone that used the shields seemed to understand their shielding value, so we decided to find a local manufacturer to see if I could start making and selling them to those who were interested. My goal was not actually to start a business, but to make these shields available for people so they could protect themselves. I did not expect them to start selling at the pace they did, and I soon needed help from my family to keep up with managing the orders. Eventually, my son Ryan left his career in New York to work with me full-time, and we have grown the business into a small company called DefenderShield®. Since then, we have expanded our product lines to include EMF-shielding cases for cell phones and tablets as well as travel pouches. Being able help people protect themselves from EMF radiation emissions has been incredibly gratifying.

Meanwhile, the scientific community's understanding of the health effects of EMF radiation continues to evolve. In fact, it has changed rapidly just within the past five years. New studies shedding vital light on the subject are being released almost on a daily basis. Research has linked EMF radiation exposure to everything from minor pains, such as headaches, to very serious concerns, such as fertility problems and cancer. For example, males who used laptops for only a few hours showed reduced sperm count, women who placed cell phones in their bras developed tumors in the breasts, and regular cell phone users have developed rare forms of brain tumors. This is just the tip of the iceberg.

Over the past few years, we noticed there wasn't one definitive source for information regarding the EMF radiation safety issue. That might even be the reason you found this book. You want answers! There just aren't any clear and concise books that lay out the whole problem and what you need to do to fix it. Well, we are changing that. Almost immediately when we began this journey, we became a resource for people who had nowhere else to turn. I have personally spent countless hours on the phone with customers, answering all types of questions. We have now helped thousands of people around the globe, while building our knowledge base of understanding. Our path led us here, and now we are going to help you.

We have written *Radiation Nation* to give you an entire overview of the EMF radiation issue. Each chapter will give you new insights so you can make wise choices when it comes to your health. We will share what we have learned and what the research is showing. You will come to understand exactly what EMF radiation is and how it affects our bodies. Based on the latest scientific studies, we will show you some of the greatest health risks and you will begin to see why the current government standards are not a good standard for safety. Finally, we will tell you the major sources of EMF emissions and the ones that are of greatest concern. This is not information that you want to push off learning until a later date. Now is the time to take control by discovering simple and practical ways you can protect yourself and your family right away.

—Daniel T. DeBaun

FOREWORD

I recently took part in an event with several chief technology officers of some of the world's top virtual reality companies and had a chance to ask them a question. In a sincere, professional way, I asked, "Whose job is it to make sure that this new VR technology is not harming our brains or screwing up our biology?"

The scary thing is, they all looked at each other and basically said, "Not ours."

We are creating amazing, world-changing technologies, and that's a good thing. I even helped to build some of them in Silicon Valley! But I wonder who will consider the potential downsides before we have deployed a billion things that can affect our brains at the cellular level?

Thousands of laboratory studies with animals and cell samples have found harmful biological effects from short-term exposure to low-intensity electromagnetic radiation (EMF) emitted from technology like cell phones and laptops. The effects include development of stress proteins, micronuclei, free radicals, DNA breakage, and sperm damage. Human studies have also found that brief exposure to cell phone radiation alters brain activity and can open the blood-brain barrier, which could enable chemical toxins in the circulatory system to penetrate the brain.

Some of these issues stem from reductions in the function of mitochondrial DNA, which are the power plants in the cell. To maintain a healthy body and live a long life, we must keep our mitochondria strong. What's happening is that when cells are constantly exposed to unnatural EMF radiation, it reduces mitochondrial function. Sperm has some of the most athletic molecules in a male's body and has the most need for mitochondria. Sperm are like little rocket ships whose job is to get to the egg as fast as possible. If there is even a small decline in mitochondrial function, some of them are not going to get there. That's one reason about 25 percent of sperm become immobile after just a few hours of exposure to EMF radiation.

The idea behind biohacking, the core principle underlying my work, is that when you change the environment outside and inside of yourself, you can control your biology. In my first book, written with my wife, called *The Better Baby Book*, we talked about fertility. We wrote that there was pretty clear evidence that you don't want to put your cell phone in your front pocket and you don't want to put your laptop right over your reproductive organs.

I am so concerned about EMF radiation that I'm taking steps to reduce unnecessary exposure in my own life, even though I use wireless devices regularly. I now have a rule in my house that my kids do not handle a cell phone or a tablet if it is not in airplane mode. They do not get to use them for hours and hours. Occasionally they can talk on the cell phone with Grandma and Grandpa using speakerphone, but they don't have to hold the phone to their head to use it. My kids will never hold the phone up to their head, and neither will I, because it is just not a good idea. I always use the speakerphone or a headset, and I don't keep my phone near my organs—especially the recreational ones! In my home, I also turn off the Wi-Fi and prefer Ethernet cabling. I have even purchased EMF-blocking DefenderPads for Bulletproof employees so they can more safely use their laptops.

I live this stuff, because our environment matters. The electromagnetic fields we are exposed to are a part of our environment that is proven without a doubt to affect our biology, even if we don't know all the mechanisms yet.

Who knows, maybe we will find out that EMF radiation is less harmful than I think it is. But is it harmless? Absolutely not. There's no evidence that it is harmless and denying the science will not help. When we admit there is a problem, we can discuss it and start hacking it. Thanks to Daniel and Ryan for bringing attention to these questions.

—Dave Asprey, founder and CEO of Bulletproof

Introduction

WHY THIS MATTERS

"Very recently, new research is suggesting that nearly all the human plagues which emerged in the twentieth century, like common acute lymphoblastic leukemia in children, female breast cancer, malignant melanoma and asthma, can be tied to some facet of our use of electricity. There is an urgent need for governments and individuals to take steps to minimize community and personal EMF exposures."

–Dr. Samuel Milham, MD, MPH, medical epidemiologist specializing in the health effects of EMF radiation

An artist depicts what Washington, DC, may look like if EMF radiation was visible to the naked eye. Photo: Nickolay Lamm.

It's Everywhere

Our world is enveloped by electromagnetic radiation, better known as EMF radiation. Every electronic device we own, from our cell phone to our laptop, emits forms of EMF. Each waking minute of every day, we surround ourselves with gadgets and gizmos that silently emit EMFs invisible to the naked eye. And we rarely stop to think about the sheer dependence we have on our electronic devices. We are too preoccupied with our busy lives to notice.

When we wake up, we hit our alarm clocks and check our cell phones. When we get to work, we power on our desktop computers or laptops. When we get home after a long day, we click on the TV with our remote control. We read with our tablets in our laps and our smartwatches strapped to our wrists. Our lives are absorbed by the new electronic age.

Most would agree that technology has improved our lives, though. Certainly, the advent of the supercomputer and microprocessor has enabled us to do things faster and more efficiently than ever before. That's why technology has proliferated. Electronics are tools that have made our lives so much more convenient.

Reasons to Care

In the pursuit for faster speed, greater productivity, and instant connection, have we neglected to fully evaluate the potential harm that we may be causing ourselves? Unprecedented levels of EMF radiation in the form of non-ionizing radio frequency (RF) and extremely low frequency (ELF) radiation surround us. The dosage is long-term and persistent.

For a long time, it was thought that only high doses of ionizing radiation like X-rays could cause serious changes to living tissue that were linked to cancer and multiple other disease states. Today, however, more recent data suggests that long-term exposure to even low levels of non-ionizing radiation emitted by our electronic devices can be harmful. Certainly, this raises the questions that public health policy experts have begun to ask. How harmful is it? Should we be concerned? Should the public be educated? This topic has become of such interest of late, it begs for more discussion in the public domain.

The Need for Better Safety Standards

Are we clueless about our EMF safety measures? Standards limits have not been updated since the mid-1990s. Photo: © 1995 Paramount Pictures.

Current EMF safety standards are limited, out-of-date, and inconsistent. In the United States, for instance, the Federal Communications Commission (FCC) regulates exposure limits for non-ionizing EMF radiation. When cell phones first appeared over three decades ago, the agency developed very basic safety standards for cell phone handling. Those standards have not been updated since 1996, even though cell phone technology has grown in power and complexity.[1]

It's not just limited to the United States, though. In all countries, there is also weakness regarding how electromagnetic radiation safety levels are set. Safety standards have only been developed with consideration to short-term thermal effects of EMF radiation on the body. The biological implications of EMF radiation exposure have not been considered or adequately measured.

Considering the upcoming 5G (fifth generation wireless frequencies of 28 GHz and above) extremely high-frequency mobile network rollout, current standards will become even more irrelevant. The current standards do not even come close to addressing 5G.

The Public's Need to Know

Asking questions is every person's right. It is reasonable to want to know what information is available today, which questions have already been answered, and which questions remain unanswered. Safety cannot be assumed, especially when EMF radiation has been shown to have a negative effect on living cells.

Unfortunately, not enough clear information has been written on this important topic for the general public to understand. We will review some of the recent findings and summarize the issue as a whole. There are numerous pointers to additional research, and both sides of the argument will be presented. We hope the reader will be encouraged to do some of their own research as a result.

A Hot Debate – Controversy and Confusion

The potential dangers of EMF radiation have become a hot debate.
Image: R. DeBaun.

The study of EMF radiation health risks over the years has been conflicting to say the least. Overall, independent studies have tended to find more of link between EMF radiation exposure and health issues. Not surprisingly, industry-funded studies have found little or no link. In the scientific forum, the validity of research efforts has been heavily contested on both sides of the issue. The legitimacy of studies is frequently called into question as well, and the scientists involved are often accused of having biases.

Over the years, the media have begun discussing the topic of EMF radiation with more frequency and seems to be as puzzled as the scientific community. Respected publications like the *New York Times* and *Forbes* have come out with a variety of opposing viewpoints. For example, the *New York Times* published an article by Kenneth Chang called "Debate Continues on Hazards of Electromagnetic Waves," in which the author described the possible adverse health effects of EMF radiation,[2] only to be countered by another writer, Geoffrey Kabat, in *Forbes*. Kabat accuses Chang's article of being outdated and only trying to raise alarm.[3]

In a more controversial article called "Could Wearable Computers Be as Harmful as Cigarettes?", also published in the *New York Times*, tech writer Nick Bilton wrote about the safety of wireless devices like cell phones and smartwatches. In the opinion piece, Bilton suggested that cell phone and Wi-Fi radiation could be viewed as dangerous as cigarette smoke was in the '70s. He also pointed to studies and cited experts that suggested cell phone radiation could be carcinogenic. Based on his research, Bilton concluded, "I have realized the dangers of cellphones when used for extended periods, and as a result I have stopped holding my phone next to my head and instead use a headset." He went on to say, "When it comes to wearable computers, I'll still buy the Apple Watch, but I won't let it go anywhere near my head. And I definitely won't let any children I know play with it for extended periods of time."[4]

While his column received support from several media outlets, it also received ridicule from sources across the media landscape despite the evidence given by Bilton.[5] The New

York Times even distanced itself from the piece by appending an "Editor's Note," in which it discredited Bilton's stance by saying the article "gave an inadequate account of the status of research about cell phone radiation and cancer risk. Neither epidemiological nor laboratory studies have found reliable evidence of such risks."[6] They also went as far as to change the article's title to the less controversial "The Health Concerns in Wearable Tech."

In both examples, we saw conflicting messages coming from the same source. How is the common reader to make heads or tails out of the information?

The Bottom Line

The authors of this work believe that EMF radiation exposure is a public health concern. We believe the evidence available today is sufficient to draw this conclusion; so much so that we have spent a lot of our time, energy, and resources to gather evidence in one place and produce this book. The information provided here originates from a variety of sources, including scientific publications, news articles, and multimedia sources. We have interpreted the evidence to mean that long-term exposure to low levels of EMF radiation has the potential for many health implications.

Nevertheless, we want you to understand that we are not anti-technology, and we do not want you to throw away your mobile devices. Far from it! We are technology users ourselves, and we are not going to give up our electronic toys anytime soon. This book is meant to help educate and offer guidance so you can take the necessary precautions to use and enjoy technology in the safest way possible. We'll be exploring what EMF radiation is, its health risks, and what you can do to protect yourself.

Science is still exploring this topic, and it will be science that we use in our discussion. There is no magic here—just logic, physics, and reputable scientific studies. Because these technologies are here to stay, we should try to better understand how they fit into our daily lives.

Chapter 1
OUR WIRELESS WORLD

"The adverse effects of electrosmog may take decades to be appreciated, although some, like carcinogenicity, are already starting to surface. This gigantic experiment on our children and grandchildren could result in massive damage to mind and body with the potential to produce a disaster of unprecedented proportions, unless proper precautions are immediately implemented."

–Dr. Paul J. Rosch, MD, clinical professor of medicine and psychiatry, New York Medical College

"Joel, this is Marty. I'm calling you from a cell phone, a real, handheld, portable cell phone."[7]

Cell phone inventor Martin Cooper makes the first mobile phone call in 1973. Photo: Courtesy of Martin Cooper.[8]

On April 3, 1973, Martin Cooper, the chief engineer at Motorola and the inventor of the cell phone,[9] spoke these words. Inspired by Dick Tracy's two-way radio wristwatch,[10] the cell phone is perhaps one of the greatest inventions of the century. Cooper's brainchild marked a technological breakthrough that was to change how the world communicates and has had even a greater impact than perhaps even he could have foreseen.

While being interviewed by reporters in New York, Cooper made the first public call ever from a handheld mobile phone to Dr. Joel S. Engel, his chief rival at Bell Labs.[11]

The call was made for publicity (and some good-natured gloating), yet it heralded the birth of a new industry and led to years of rapid technological change.

It took nearly a decade for Cooper's invention to fully transform from science fiction to reality when Motorola released the first commercially available mobile phone, the DynaTAC 8000x, in 1983. Throughout the 1980s and early 1990s, mobile phones slowly gained acceptance, mostly among businessmen. Yet it wasn't until the late 1990s and 2000s that cell phones truly gained mass appeal.

How Did We Get Here?

Just a few decades ago, computers took up entire floors, but seemingly overnight, we are carrying mini computers in our pockets and purses as a matter of course. Advances in technology exploded over an extremely short period of time, and nobody—not even the telephone companies—was prepared.

Back in the '80s, I (Daniel) was in an executive meeting at a leading telecommunication company and we were discussing what the "next big thing" in the telephone industry would be. They were betting the farm on serving customers who wanted second phone lines, or "princess phones."

My quick reaction was, "What about cell phones?"

Their response? "We don't think revenues will be in wireless service." With that skepticism and summary dismissal, they forfeited the prize of leading the market with wireless technology.

Some companies, like Motorola and other small start-up firms, were not so hasty to ignore this technology. Today, of course, all of the small businesses that dove headfirst into developing wireless technology have been bought up by the large telecom companies. The big telecoms may have been late to the game with wireless, but they recovered by sucking up the competition. Now, it's their landline-phone business that's nearly extinct.

In 2005, only 33.9 percent of people across the world had a mobile subscription. By 2015, that number almost tripled to 96.8 percent![12] In 2016, there were over 7 billion mobile subscriptions worldwide.[13] Along with this growth, over the past decade, we've also seen the evolution from flip phones to smartphones, such as the iPhone, which have paired computer technology with cellular capability. In addition, we are also able to connect to the Internet and other devices via wireless technologies like Wi-Fi and Bluetooth. Bill Gates's goal of putting a computer in every home seems quaint in a world where the growing trend is a computer in every pocket! We now hold extraordinarily powerful computers right in the palms of our hands.

Saved By the Bell, a popular show in the late 1980s and early 1990s, features Mark-Paul Gosselaar as Zack Morris talking on his "brick" cell phone. Photo: NBC Productions.

Health Concerns Emerge

Unfortunately, increased computing capacity and connectivity create a downside. All cell phones emit invisible low-frequency radiation called electromagnetic radiation (EMF radiation). Low-level radiation given off by cell phones has been generally considered "safe" to humans, but we are now beginning to learn that this may not be the case at all. Exposure to EMF radiation could be a source of underestimated health risks; however, it's surprising that so few are even aware of the dangers, much less the whole issue.

Because cell phone use grew so rapidly, possible health concerns initially took a back seat. However, now that their undeniable popularity has been completely recognized, greater interest in the effects of their emissions has grown. To date, numerous scientific studies have investigated the possible correlations between mobile phone radiation exposure and illness. Most of these studies have been either short-term or have used small sample sizes, though, so their findings have been heavily interpreted and hotly debated. While all scientists agree that more study is needed, researchers are still trying prove or disprove a causal relationship.

By 2011, enough data had been accumulated to show that some risk existed due to long-term, heavy use of mobile phones, compelling the International Agency for Research on Cancer, a branch of the World Health Organization, to classify mobile cell phone radiation in the Group 2B category, indicating a possible carcinogen (a substance or source of exposure that can cause cancer).[14] This is the same category as DDT, lead, engine exhaust, chloroform, and glyphosate (the active ingredient in Roundup®). Later, in 2016, a $25 million study conducted by the National Toxicology Program (NTP), part of the National Institutes of Health, confirmed what many have believed for years—that exposure to EMF radiation emitted from cell phones can lead to serious health issues including brain and heart tumors.

A Sea of Electronics

The problem isn't just from cell phones, though. Everywhere you look, there is new technology carrying potential health impacts that we may have never considered before. Desktop computers, laptops, tablets, and other mobile devices all emit forms of EMF radiation. Today, these devices are in our homes, schools, workplaces, cafés, and public areas—and they are only growing in popularity.

According to data from Pew Research, as of mid-2015:

- 92 percent of US adults own a cell phone (including smartphones)
- 71 percent of US adults own a desktop or laptop computer
- 45 percent of US adults own a tablet computer

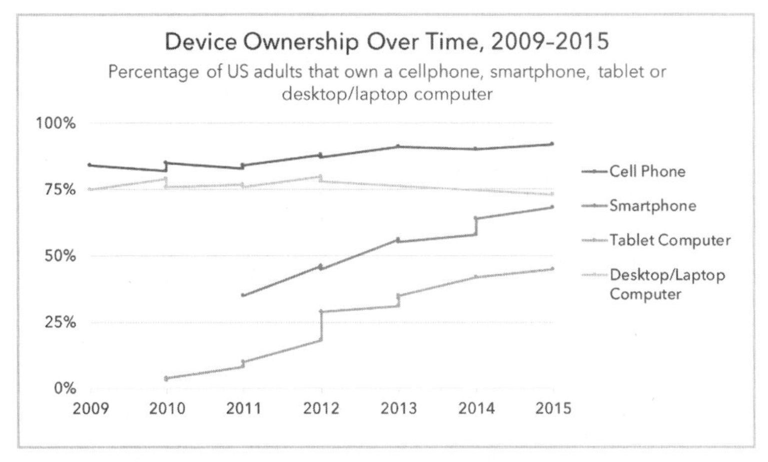

Data Source: Pew Research Center.[15] *Image: R. DeBaun.*

Not only is the use of electronic devices growing, but the time spent using them is increasing also, especially in the United States. Young and old, we love our electronics. In 2016, *The Total Audience Report* released by Nielsen showed the average American spent nine or more hours a day using electronic media.[16] Given that the average human spends seven to nine hours sleeping each night, that means that we spend around two-thirds of our waking hours "wired." That adds up to a huge percentage of a person's lifetime. Think of the dosages we are exposed to during the course of our day *and* night.

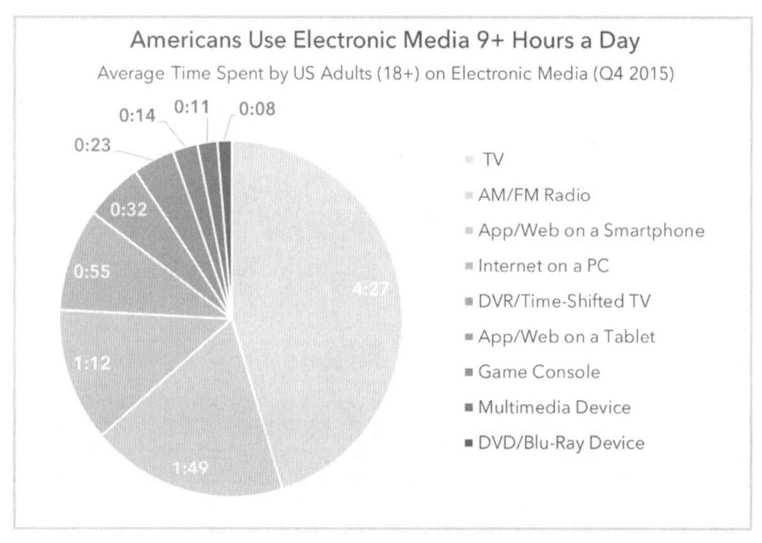

Data Source: The Nielsen Company.[17] Image: R. DeBaun.

A Global Wake-up Call?

Around the world, people are beginning to realize we may have a problem and are taking action. Many countries have issued precautionary health warnings about exposure to EMF radiation, and they are making recommendations on how to reduce the risks. In the United States, despite the telecom industry arguing against precautionary warnings, states, city governments, and institutions are starting to pass laws warning of the health risks.

Following are actions being taken around the world:

- The city of Berkeley, California, has enacted an ordinance requiring cell phone labeling to warn of the health risks from radio frequency (RF) radiation.[18]
- The French National Assembly recently passed a law that will limit young children's exposure to EMFs generated by wireless technologies. The law includes bans of Wi-Fi in any childcare facilities that cater to children under the age of three; it also requires that cell phone manufacturers recommend the use of hands-free kits to everyone. Additionally, the law also bans any advertising that specifically targets youth under the age of fourteen.[19]
- The National Library of France has banned Wi-Fi.[20]
- The Italian State Parliament of South Tyrol mandated that the state government replace existing wireless networks with networks that emit less radiation at schools, hospitals, nursing homes, and other public facilities whenever possible. It also voted to start an education and awareness campaign that informs the public about possible health risks, especially with regard to the unborn, infants, children, and adolescents, while simultaneously developing guidelines for a safer use of cell phones, smartphones, and Wi-Fi.[21]

- The city of Mumbai, India, has adopted a policy prohibiting cell towers in schools, colleges, hospitals, orphanages, and juvenile correction homes. It also prohibits nearby antennas being directed toward such buildings and requires that antennas on such buildings be removed.[22]
- The Israeli Ministry of Health has issued recommendations and has initiated a major public awareness effort to reduce wireless radiation exposure to children.[23]
- In Denmark, the European Environment Agency released a review entitled "Late Lessons from Early Warnings: Towards Realism and Precaution with EMF?" In this review, some of the debates on the safety levels of EMF are discussed, and risks and strengths of various pieces of evidence are described. Ultimately, the review urges the public to help participate in risk analysis as our scientific body of knowledge on EMF grows.[24]
- The Elementary Teachers Federation of Ontario, a union representing seventy-six thousand teachers, called for moratorium on the use of wireless in schools. Andrea Loken, president of the Ontario Secondary School Teachers' Federation Limestone District, stated, "We've never been asked if we're OK with being subjected to Wi-Fi all day everyday while we're at work. No one

has given consent and no one has been informed of the risks."[25]

- Lakehead University in Canada has banned Wi-Fi on campus.[26]

These actions are just a sample, but it seems they are gaining some momentum globally. Ironically, many of the world experts on EMF are from US institutions and universities. The dangers of EMFs are surfacing in mainstream media yet not enough for the health risks to be common knowledge. Americans aren't fully informed and don't realize that there may be a need for policy changes, nor do they know the proper actions to take for their own safety.

As this information is slowly being released, many citizens are taking the bull by the horns and forming organizations to discuss the issues. Although activism is generally a good thing, the problem with these efforts is that the information being disseminated is not necessarily filtered through any scientific lens. Much of the information is helpful, but not all is accurate. Misinformation can be as bad as lack of information, because it undermines proper education on the subject. More efforts need to be made by government institutions and research centers to disseminate the correct information.

Most consumers are still not aware that EMF radiation emissions, from such innocuous-looking devices as cell phones and tablets, can be harmful. For example, toys are currently being marketed as kid-safe, despite the fact they produce radiation that children are the most susceptible to. Parents freely expose their children to radiation despite the fact that children's soft tissue is still developing.

An Ongoing Discussion

Over the years, there has been heavy discussion about whether EMF radiation is harmful. Whenever a controversial topic is new, there is certain to be arguments on both sides of the issue. However, the winds are beginning to shift with new information being discovered. In recent years, a significant and growing body of evidence is showing that radiation from electronic devices is not completely safe. When variations are introduced into our environment, such as the introduction of cell phones, potential health issues may emerge quickly, but problems can take many years—or decades—to appear. With EMF exposure, it has taken some time, but we are now seeing a pattern of problems similar to what we witnessed with cigarette smoking.

Dr. Joel Moskowitz, PhD, director of the Center for Family and Community Health at the University of California, Berkeley, puts it this way: "Very few health scientists or medical professionals argue either that cell phones are 'absolutely safe' or 'absolutely dangerous.' Most now recognize that cell phone radiation is biologically active and agree we need more research to determine in what ways it can be harmful to our bodies and how best to reduce risks. Many scientists and health professionals recommend precautionary measures to the public. Some others who have conflicts of interest or interpret the data more conservatively tell their family members to take precaution."[27]

Our Children

What is worse than the proliferation of radiation-emitting devices surrounding us every day is how our children are being exposed. As anyone over the age of thirty knows, we didn't grow up using cell phones, tablets, laptops, and other electronics against our bodies every day. But we now live in a different world.

April 4, 2005: Pope John Paul II's body is carried into St. Peter's Basilica for public viewing. Photo: Luca Bruno/AP.[28]

March 13, 2013: Pope Francis makes his inaugural appearance on St. Peter's balcony. Photo: Michael Sohn/AP.[29]

Today, kids start using these devices when they are barely toddlers. Children are particularly vulnerable to EMF radiation because of their developing bodies. With exposure beginning at such a young age, and sure to last throughout a lifetime, we consider children to be at the most risk. To bring focus to this concern, we devote part of chapter five to helping parents and children understand the health risks using modern electronic devices.

We hope that these issues of EMF radiation exposure will help everyone become more informed.

Chapter 2

WHAT IS EMF RADIATION?

"Any possible health effects of ELF-EMFs would be of concern because power lines and electrical appliances are present everywhere in modern life, and people are constantly encountering these fields, both in their homes and in certain workplaces. Also, the presence of ELF-EMFs in homes means that children are exposed. Even if ELF-EMFs were to increase an individual's risk of disease only slightly, widespread exposure to ELF-EMFs could translate to meaningful increased risks at the population level."

–National Cancer Institute

Photo: R. DeBaun.

What do people even agree about? Well, we all seem to agree that EMF radiation surrounds all of us. There are countless natural and unnatural sources of EMF radiation. The sun, the stars, and the sky are all natural forms. TVs, hair dryers, alarm clocks, microwaves, computers, tablets, and cell phones are all examples of human-made sources. We may not always be able to see or feel it, but we can all generally agree that EMF radiation is everywhere, for better or worse. We can also all agree that we are being dosed at low levels of radiation over long periods of time. Where we may disagree is how detrimental this radiation exposure actually is to us.

Understanding what EMF radiation is, in *all* its forms, can help you deal with its unavoidable presence in modern life, as well as help you lessen your risks of any unwanted long-term exposure.

EMF Radiation Defined

EMF radiation is *energy* that travels through space in electromagnetic waves such as radio waves and light waves. This energy is made up of small packets of particles called photons that can either travel alone or move around together in synchrony.

The distance between these energy waves is called a *wavelength*. You can envision these as the waves on an ocean, being measured by the distance from one crest to the next.

Frequency can be described as the number of wave crests moving past a certain point in a given time frame.

How powerful a type of EMF radiation is comes down to its wavelength and frequency. For instance, radio waves have much longer wavelengths and slower frequencies than light waves, which have shorter wavelengths yet move at a faster frequency.

All levels of EMF radiation fall somewhere within the electromagnetic spectrum, which can be visualized like so:

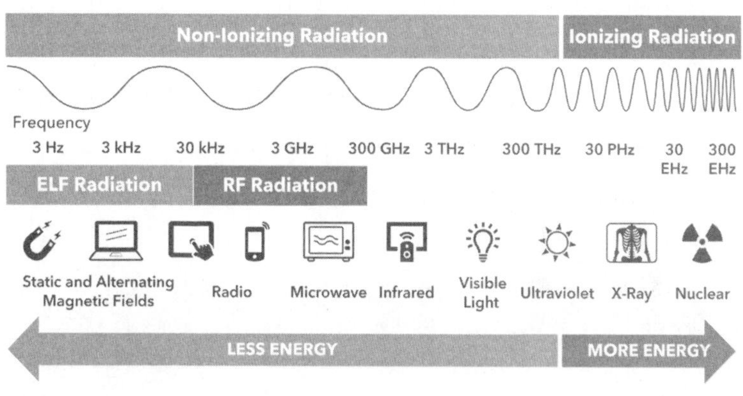

Image: R. DeBaun.

Ionizing Radiation vs. Non-Ionizing Radiation

Radiation is also separated into another two groups: ionizing radiation and non-ionizing radiation. It is worth noting that there is a difference between how ionizing radiation and non-ionizing radiation is viewed in relation to health risks.

Ionizing Radiation

Ionizing radiation has waves with short wavelengths that move at a high frequency. It is the type of emission considered to be the most dangerous, because it can cause permanent damage to the body and even kill with a single exposure. X-rays, gamma rays, and nuclear bombs produce ionizing radiation. Ionizing radiation is capable of rupturing atomic structures and even fracturing the molecules in our bodies instantly.

Non-Ionizing Radiation

Non-ionizing radiation moves at a low frequency with a long wavelength and is weaker than ionizing radiation per dose. While it's true that the energy of non-ionizing radiation does not fracture a molecule in the same way as ionizing radiation can, recent evidence demonstrates that non-ionizing radiation can alter cells on a biological level.[30] This is significant because even though you may not be able to see it, non-ionizing radiation exposure can eventually impact the cells of your body, particularly over a long period of time.

Types of EMF Radiation Emitted from Electronic Devices

Electronic devices emit two forms of non-ionizing EMF: extremely low frequency (ELF) and radio frequency (RF) radiation.

ELF Radiation

Electronic devices all work pretty much the same way. They use an electric current that passes through a circuit to perform a function. Electricity is a form of energy that is used to make the device do what we want. When currents flow within and between components in an electronic device, it will emit a form of EMF called ELF radiation. ELF radiation is measured in gauss (G) or milligauss (mG), which is 1/1000 of a gauss. ELF radiation oscillates at a frequency below 300 hertz (Hz), which is on the very low end of the electromagnetic spectrum.

RF Radiation

When electronic devices, such as laptops and cell phones, connect to the Internet via Wi-Fi or cellular signals, they emit an additional form of EMF called RF radiation. RF signals, which are measured in volt per meter (V/m), are transmitted at many different frequencies and power levels. A cell phone signal, for example, can travel several miles, while a Wi-Fi signal stays within a few hundred feet. RF signals range between 30 kilohertz (kHz) to 300 gigahertz (GHz),[31] but cell phone signals and Wi-Fi signals oscillate between 800 megahertz (MHz) to 5.8 GHz on the electromagnetic spectrum.

Health Implications of Non-Ionizing Radiation Exposure

More and more research is revealing that exposure to non-ionizing ELF and RF radiation may cause serious health issues. We have been slow to recognize these dangers because the biological impacts of non-ionizing radiation take time to develop; it's hard to spot danger when it advances at a snail's pace. For example, if you shine a high-energy laser beam on a penny, it will melt right away. However, if you shine a low-energy beam on a stone, nothing immediately happens. Over time, though, you could potentially carve a hole in the stone.

Non-Ionizing Radiation Mechanism of Action

It is well-known that non-ionizing radiation can cause heating of cells,[32] yet evidence is now showing that it may also cause cellular changes by opening voltage-gated calcium channels. According to Dr. Martin Pall, PhD, non-ionizing radiation exposure disrupts the normal flow of calcium in the body, starting a cascade of events leading to cancer.[33] Another recent study by Dr. Nora Volkow, MD, indicates that when a cell phone is used for only fifty minutes, brain tissues on the same side of the head as the phone's antenna metabolize more glucose than tissues on the opposite side.[34] However, the long-term effect of this is still unknown.

Although studies linking the health effects of long-term non-ionizing EMF exposure may take years to reach definitive conclusions, following is a list of the risks emerging:

- Reproductive effects (e.g., reduced sperm count and sperm motility)
- Genotoxic effects (e.g., DNA fragmentation, fertility problems)
- Glioma and meningioma (forms of brain cancer)
- Cellular stress, including release of heat shock proteins
- Cognitive effects (e.g., lethargic response times)

- Electromagnetic Hypersensitivity Syndrome, or EHS (e.g., tingling sensations; fatigue; sleep disturbances; dizziness; loss of mental attention, reaction time, and memory retentiveness)
- Behavioral effects
- Integrity changes to the blood-brain barrier
- Sleep and electroencephalogram (EEG) effects

This is an extensive list, and the data is hard to ignore, yet it is only a sample. We will delve further into the health risks of EMF radiation in later chapters.

Chapter 3

HOW EMF EXPOSURE AFFECTS THE BODY

"Cells in the body react to EMFs as potentially harmful, just like to other environmental toxins including heavy metals and toxic chemicals. The DNA in living cells recognizes electromagnetic fields at very low levels of exposure, and produces a biochemical stress response."[35]

–Dr. Reba Goodman, PhD, professor emeritus, clinical pathology, Columbia University

"The scientific evidence tells us that our safety standards are inadequate, and that we must protect ourselves from exposure to EMF due to power lines, cell phones and the like, or risk the known consequences. The science is very strong and we should sit up and pay attention."[36]

–Dr. Martin Blank, PhD, Department of Physiology and Cellular Biophysics, Columbia University

Image: R. DeBaun.

The human body is a beautifully complex and fragile masterpiece. Each of us is an organism made up of millions of cells, working together in perfect balance and harmony. These cells are comprised of delicate molecular machinery, such as proteins and DNA, and are influenced by the world around them. Cells must work together and do so by communicating with one another through electrochemical means. When we are healthy, the signals in our body are functioning correctly to achieve a common goal. When we are unhealthy, those signals are thrown off-balance and result in sickness or worse. The perfect dance between life and death, health and illness, comes down to how our cells react to their environment.

Cell Communication

A series of electrical impulses control the human body. Our nerves detect pain or pleasure and send a message to the brain through an electric signal; the brain then shoots back a command. The trillions of cells within us are coordinated in movement and response because electrical signals are converted to chemical ones. These signals are cascaded from one cell to the next, so that one cell can send a message to another and the individuals can respond as one. This is how our body functions and all biological life on earth exists.

Our cells communicate through signals to achieve a purpose, so their ability to perceive and respond to their environment is the crux of the body's ability to function. Cells sense their environment and react. For example, the specialized cells of the nervous system (called neurons) are designed to carry messages from cells to the brain. When the neuron is sufficiently stimulated (maybe you stepped on a nail), it "fires" and you feel pain. Even though neurons are specialized messengers for the brain, all cells possess energy capable of influence by an electric field. Since EMF radiation surrounds us, it only makes sense that it can unduly alter the behavior of our cells. This can lead to abnormal changes within the cells, including irreparable damage and/or undesirable outcomes.

In fact, researchers at Tufts University have empirically proven that cellular growth and development is sensitive to bioelectrical communication. If an outside source disrupts a cell's normal electrical potential, it is possible for the source to harm the cell, or to influence its growth in an abnormal way (e.g., force tadpoles to grow eyeballs on their tails).[37]

Self-Defense Mechanisms of the Cell

Research by Dr. Martin Blank and Dr. Reba Goodman of the Department of Physiology at Columbia University has shown that EMF radiation exposure activates a cellular protective mechanism called a stress response. The stress response increases levels of specific stress proteins, called 70 Kilodalton Heat Shock Proteins (HSP70s). These HSP70s help chaperone damaged proteins by transporting them across membranes for cellular repair.

Generally, in order to activate stress response in genes and proteins, specific breaks in DNA strands must occur first. Cells monitor these DNA breaks. According to Blank and Goodman's findings, EMF radiation can directly interact with electrons in DNA, stimulating stress. The cell gets confused and produces a response indicative of damage. The researchers concluded that we should change our current EMF safety limits to account for these effects on cells.[38]

In a 2000 Hungarian study, cellular responses to various forms of EMF radiation were documented. EMF radiation was found to cause either reversible and irreversible structural and functional changes to cell organelles, which are the specialized subunits within the cells that allow them to function. The study found that reversible alterations occurred in the structure of cell organelles. However, perhaps more troubling, morphological signals associated with cell death were also triggered.[39]

A 2005 independent Italian study corroborated the Hungarian findings. It found that a cell phone frequency induced apoptosis, or programmed cell death, in human recombinant DNA (artificial DNA created by combining DNA from different organisms or cells). It also found that radiation-induced damage included changes in the organization of the cell membranes. Virtually every important subunit, including the mitochondria, endoplasmic reticulum, Golgi complex, and lysosome system, was affected in some way by EMF exposure. When viewed through an electron microscope, changes in the appearance of the cell could also be seen. The researchers believed that this was the result of radiation causing damage to DNA or proteins in the nucleus, or "brain," of the cell.[40]

Counterarguments have said that the damage was not permanent and was only part of a stress response, but we need to pay attention to the study of stress proteins. As Dr.

Leif Salford of Lund University in Sweden puts it, "Molecules such as proteins and toxins can pass out of the blood, while the phone is switched on, and enter the brain. We need to bear in mind diseases such as MS and Alzheimer's are linked to proteins being found in the brain. Because stress proteins are involved in the progression of a number of diseases, heavy daily cell-phone usage could lead to great incidence of disorders such as Alzheimer's and cancer."[41]

Overwhelming Our Defenses

Our bodies are not without defense mechanisms. If our body experiences a small amount of damage, generally that is not too bad and we can cope. We have mechanisms in our body to repair such things as breaks or mutations in our DNA. Occasional breaks in DNA are actually normal and can happen through routine exposure to the environment (such as from UV radiation from the sun). However, if we receive too much damage, it may become difficult for our body's defenses to cope with overwhelming problems. Thus, permanent and irreparable changes to a cell can lead to disease states.[42]

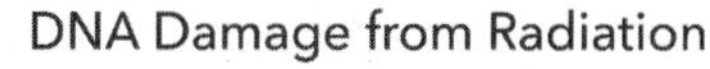

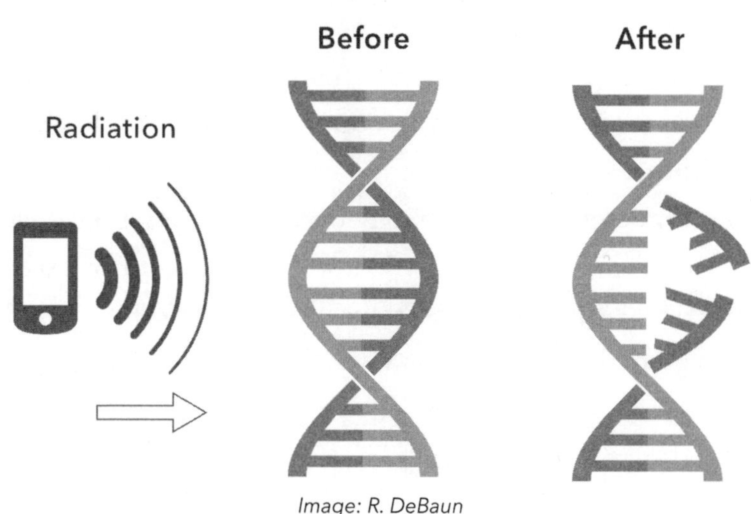

Image: R. DeBaun

According to the Centers for Disease Control and Prevention (CDC), "Workers exposed to high [electromagnetic] fields have increased cancer rates."[43] Although the CDC does not necessarily conclude that EMF exposure causes cancer, it does note the correlation. In the case of cancer, a damaged cell, not fixed or destroyed, can proliferate out of control and become a tumor or growth.

Effects of EMF on Brain Function

Recent research led by Dr. Volkow has shown that just under an hour of cell phone use is linked to an increase in brain activity in the area closest to the phone's antenna.

Researchers undertook a random study of forty-seven participants to determine whether exposure to cell phone radiation had an effect on regional activity occurring within the brains of humans. Cell phones were positioned on both the left and right ear while an image of the brain was captured using a positron emission tomography (PET) brain scan together with an injection of 18F-fluorodeoxyglucose to measure glucose metabolism in the brain. This was performed twice: first, with the cell phone on the right ear turned on for fifty minutes with the sound muted ("on" state); second, with both mobile devices turned off ("off" state).

After analyzing the data to determine the association between brain metabolism and the estimated levels of RF emissions produced by the handsets, the researchers compared the PET brain scans to determine the effect that cell phone use had on glucose metabolism in the brain.

While there was no difference in whole-brain metabolism when the phones were turned on or off, the researchers did find that there were significant differences in metabolism within regional areas of the brain. Metabolism was approximately 7 percent higher in the region of the brain that was closest to the cell phone's antenna when the devices were turned on compared to when the devices where turned off, which the authors considered significant.

According to the authors, "The increases were significantly correlated with the estimated electromagnetic field amplitudes both for absolute metabolism and normalized metabolism. This indicates that the regions expected to have the greater absorption of RF-EMFs from the cell phone exposure were the ones that showed the larger increases in glucose metabolism."

These results of this study show that human brains are sensitive to exposure from acute cell phone use; however, the mechanisms by which RF radiation could affect brain glucose metabolism are unclear.[44]

In an accompanying editorial article related to Dr. Volkow's study, Dr. Henry Lai, PhD, and Dr. Lennart Hardell, MD, PhD, noted that the increased brain activity observed in this study may just be the tip of the iceberg and could be indicative of "other alterations in brain function from Radio Frequency emissions, such as neurotransmitter and neurochemical activities. If so, this might have effects on other organs, leading to unwanted physiological responses. Further studies on biomarkers of functional brain changes from exposure to Radio Frequency radiation are definitely warranted."[45]

Why Isn't Everyone Sick?

So, you may be asking, since EMF radiation is all around us and it can affect our biology, will everyone get sick? For most of us, the answer is no, and that is comforting. *But*, it is hard to tell who will be affected and who will not, because there is a plethora of variables that come into play.

For instance, some people have strong genes. Or perhaps individuals haven't been using their electronic devices long enough to experience ill effects. Perhaps they don't even use their devices close to their bodies. These are just some reasons that some people won't experience health problems from EMF radiation. But does this mean there isn't a problem? Can some of us be experiencing issues and simply be unaware? Are we even asking the right questions?

Maybe we need to be more specific. For a moment, let's simplify the issue by ignoring all the less-severe effects (such as fatigue, headaches, and tingling sensations) and ask only one specific question: Can cell phones lead to brain cancer?

There is a growing amount of medical evidence to show an increasing trend that links one type of brain cancer to cell phone use. Unfortunately, reports on this subject can vary greatly. Some reports say there is no increasing trend in brain cancer, so therefore, there is no sickness linked to cell phones. So why the discrepancy?

There are actually many forms of brain cancers. Worldwide, incidences of brain cancer show that some types are increasing while others are decreasing; however, on the average, trends are flat. The media frequently uses this data to claim that no cell phone radiation and brain cancer linkage exists; however, there is a major point often overlooked about this flat trend. There is really only one brain cancer typically linked in medical studies to cell phone use: gliomas called glioblastomas. Therefore, to determine if there *is* a trend between brain cancer and cell phone usage—a trend to worry about—the first condition to investigate is glioblastomas.

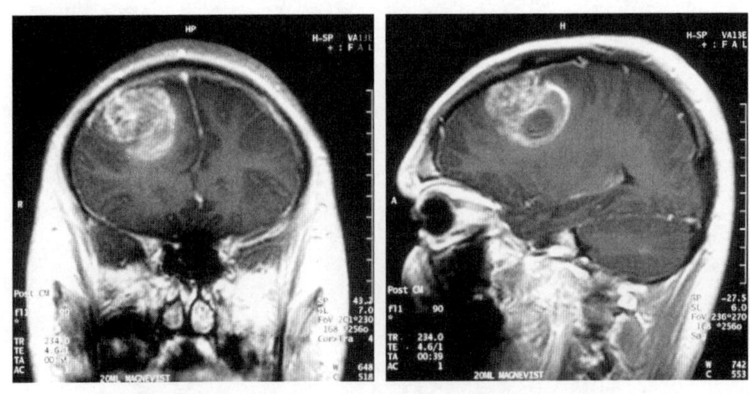

The light-colored mass in this brain is a glioblastoma.
Photo: A. Christaras via Wikimedia Commons.[46]

The average time between first exposure to a cancer-causing agent and diagnosis of the disease is 15–20 years or longer, according to the CDC.[47] Extensive use of cell phones has only occurred within the last 15–20 years, and it also takes time for science to study trends in populations. It also requires the analysis of large bodies of data over long time periods. However, research has shown an upward trend in glioblastomas located in parts of the brain proximal to where cell phones are held: the cerebellum, the frontal lobes, and the temporal lobes.[48] For example, a 2009 Australian study found that glioblastomas increased by 2.5 percent each year from 2000 to 2009 in New South Wales and the Australian Capital Territory. Another study, released in 2012, found that glioblastoma incidence in the United States increased in the frontal lobes 1.4–1.7 percent from 1999 to 2000. In the temporal lobes, it increased 0.5–0.9 percent annually.[49] This is statistically significant evidence. The following graph illustrates the correlation between cell phone use, odds of developing a brain tumor, and latency in number of years it can take to develop.

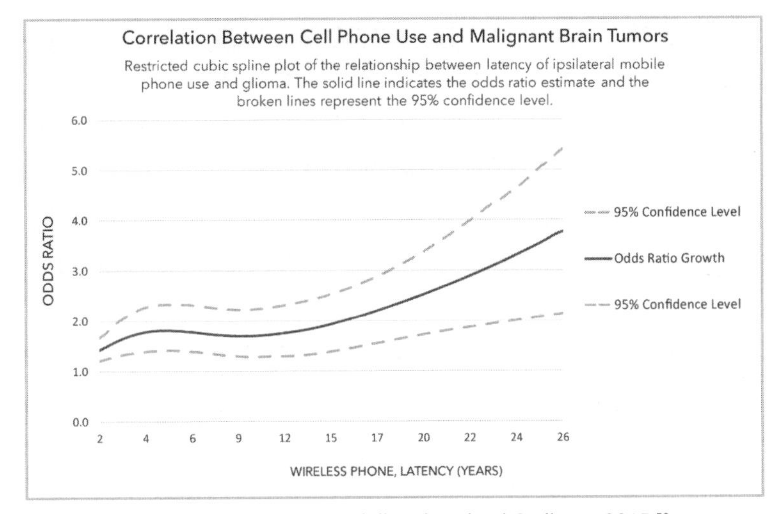

Data Source: Lennart Hardell and Michael Carlberg, 2015.[50]
Image: R. DeBaun.

As a scientific standard, it is not enough to show a higher incidence between increased cell phone use and cancer—scientists must provide evidence of exactly how the cell phone emissions *cause* brain cancer for science to call it proof. In the case of human study, this means we have to wait until a large percentage of cell phone users develop cancer to prove direct causation. This also means we must be able to eliminate or somehow ignore other risk factors, such as the role of genetics and environmental factors. With that said, a 2016 study could be the turning point in proving that EMF radiation exposure can lead to cancer.

National Toxicology Program (NTP) Study

In May 2016, perhaps the most conclusive and official evidence to date that shows a link between non-ionizing cell phone radiation and brain cancer was released (in partial findings) by the US government. In a $25 million, in-depth study, conducted by the National Toxicology Program (NTP), researchers found that rats exposed to RF radiation had higher rates of glioma (a type of brain tumor), as well as malignant schwannoma (a very rare heart tumor), than unexposed rats.[51]

In the study, which is the largest and most complex ever conducted by the NTP,[52] thousands of rats were exposed to three levels of RF radiation—1.5, 3, and 6 watts per kilogram (W/kg)—over two years. They were also exposed to radiation from two different types of mobile technology, GSM (Global System for Mobiles) and CDMA (Code Division Multiple Access). Radiation exposure level showed a direct dose-response relationship, meaning the more radiation rats received, the higher the amounts of brain and heart cancer were seen. There were no tumors found in the rats not exposed to radiation.[53]

Overall, 8.5 percent (46 out of 540) of male rats exposed to cell phone radiation developed cancer or precancerous cells. None of the ninety unexposed male rats developed cancer or precancerous cells.

In the group of rodents exposed to the lowest intensity of radiation (1.5 W/kg), 12 out of 180, or 6 percent, of male rats developed cancer or precancerous cells.

Of the rodents in the highest exposure group (6 W/kg), 13 percent (24 out of 180) male rats developed cancer or precancerous cells.[54]

In a statement released with the study, the researchers said, "Given the widespread global usage of mobile communications among users of all ages, even a very small increase in the incidence of disease resulting from exposure to [wireless radiation] could have broad implications for public health."[55]

The NTP study is extremely significant because it is considered to be accurate and well designed by leading health authorities. In fact, it is one of the largest and most in-depth analyses of mobile phones and cancers conducted to date.

RF frequencies are thermal emissions, known to heat up your body's cells.[56] As we will discuss later in the book, our belief is that the current FCC standards, which allow for cell phone RF emissions outputs of 1.6 W/kg, are insufficient. This standard is designed to limit RF radiation from penetrating an adult's head more than one inch, while not allowing for a temperature increase more than two degrees Fahrenheit in the area of the head touched by the cell phone.

The NTP study is different because it goes beyond the current FCC standards and focuses on the biological impact of radiation to the body. A distance between the RF source and the testing subject was established to ensure there were no thermal influence on the cells. Therefore, the NTP study proves that non-ionizing, RF radiation can cause cancer without heating tissue and makes a strong argument that the FCC's standards do not provide sufficient protection.

Otis W. Brawley, MD, chief medical officer for the American Cancer Society, stated about this study, "For years, the understanding of the potential risk of radiation from cell phones has been hampered by a lack of good science. This report from the National Toxicology Program (NTP) is good science. The NTP report linking radiofrequency radiation (RFR) to two types of cancer marks a paradigm shift in our understanding of radiation and cancer risk. The findings are unexpected; we wouldn't reasonably expect non-ionizing radiation to cause these tumors. This is a striking example of why serious study is so important in evaluating cancer risk. It's interesting to note that early studies on the link between lung cancer and smoking had similar resistance, since theoretical arguments at the time suggested that there could not be a link."[57]

It is important to note that cancer is not the only illness linked to EMF exposure. Small disturbances due to EMF can lead to larger consequences in a variety of ways. Ongoing research is being done across the board on other health issues, which we will delve into further in the following chapter.

Chapter 4

EMF HEALTH RISKS

"When you look at cancer development–particularly brain cancer–it takes a long time to develop. I think it is a good idea to give the public some sort of warning that long-term exposure to radiation from your cell phone could possibly cause cancer."

–Dr. Henry Lai, professor of bioengineering, University of Washington

"It is imperative that the accumulating scientific and medical evidence of the long-term ill effects of electromagnetic fields (EMF) be taken into consideration."

–Dr. Mitchell Fleisher, MD, DH, double board-certified family physician

What it Means to be "Safe"

As human beings, we get comfortable with the things that surround us. The more familiar we are, the safer we feel. If we interact with something hundreds of times a day, and it has never hurt us, then we tend to think that it can't or won't—or that the risk is minimal. It's how humans have learned to deal with the environment: there are just so many dangers in the world, it is not practical to be in a constant state of alarm.

Evolution has programmed us to prioritize dangers, and we think the things we are at ease with are safe.

However, this sense of security can be misleading; it's not appropriate for everything. Our past precedents, or the limited knowledge we have about the behavior of common objects that surround us, are only as accurate as the information we possess. For example, if we do not know how EMF radiation affects us, how can we say we are safe? How do we know that the little doses of EMF that bombard our bodies every time we put our phones to our ears, use our laptops, or plug in appliances truly has no effect on us?

History is littered with examples of commonplace things that have turned out to be more dangerous than we thought. For example:

- common knowledge that cigarettes caused lung cancer and other diseases did not occur until the 1960s;[58]
- asbestos and lead paint were used for everything from home construction to fabric before we discovered they were highly toxic and later banned them across the globe;
- fen-phen used to be a popular dietary supplement—until research showed it could cause fatal heart problems.

Of course, that doesn't mean that anything new and widely used is automatically dangerous. We should not approach everything with a sense of apprehension or fear. But if some doubt exists, or some evidence suggests that something may cause harm to us or our loved ones, we should at least take precaution.

Areas of Concern

We are surrounding ourselves with EMF-radiating devices constantly and the data clearly shows something is happening on a cellular level when we are exposed. EMF exposure has been linked to a variety of disease conditions and precursor states. Numerous studies correlate EMF exposure to everything from cancer and reproductive issues to neurobehavioral abnormalities. A cross section of the most notable examples is compiled below as a reference. This list is not fully comprehensive but is meant to give you an idea of the wide range of health concerns that have been linked to EMF exposure. Every day, our understanding develops as the research findings grow.

Asthma

A study has found that babies born to women exposed to EMF radiation during pregnancy had more than three times the risk of developing asthma. The study also showed that two known asthma risk factors—the mother having asthma and her baby being her firstborn child—exacerbated the effects of EMF exposure on asthma risk. According to the researchers, this supports the link between EMF exposure and asthma even further.[59]

Autism

Autism spectrum disorder (ASD) has been linked to the biological effects of RF radiation exposure. Evidence shows that RF exposure causes damage and cellular stress to proteins which results in neurological harm. In people who already have ASD, it makes the symptoms worse.[60] It has also been suggested that fetal or neonatal exposures to RF radiation may be associated with an increased incidence of autism.[61]

Cell Changes and Genetic Damage

Research has found that both ELF and RF radiation can cause cell changes, as well as impairment and mutations in DNA, which can lead to cancer and other health issues. Other studies find that EMF radiation may lead to changes in the physiological indices, genetics, and immune function of the operators as well as cause radical formation and direct damage to cellular macromolecules.[62]

Cancer

Cancer is probably one of the most researched areas in this field, and there have been hundreds of peer-reviewed studies conducted over the years to find out if EMF exposure increases cancer rates. Studies have been conflicting, but overall there is evidence that long-term exposure to EMF radiation may increase the risk of developing cancer. In addition, those exposed to higher levels develop it more quickly than those not exposed.

Brain Cancer

Typically, brain cancer is a rare condition, and it accounts only for 1.4 percent of all cancers and 2.4 percent of all cancer-related deaths.[63] However, as mentioned in the previous chapter, the prevalence of some forms of brain cancer is actually increasing, and several studies from the past few years point to cell phone radiation exposure as one of the possible causes.

In addition, the NTP study and other previously discussed brain-related disease research published in *Cancer Epidemiology, Biomarkers & Prevention* in 2014 showed an association between work-related exposure to ELF radiation and glioma. The researchers of the study also stated that ELF radiation could promote brain tumor formation.[64]

Also in 2014, Swedish researchers analyzed case-controlled studies of patients with malignant brain tumors diagnosed between 1997 and 2003 and 2007 and 2009. They concluded that glioma, as well as acoustic neuroma, were linked to RF radiation emissions from mobile phones. The highest risk for glioma was found in the temporal lobe; due to these high rates of risk, the researchers suggested that cell phones be regarded as carcinogenic under the Group 1 IARC classification, instead of their current Group 2B, "possible carcinogenic" status.[65]

Finally, a German study released in 2015 indicated that exposure to RF radiation levels below current safety limits still had tumor-promoting effects on mice.[66]

Breast Cancer

In 2013, researchers at the University of California published their findings on young women with prolonged contact between their breasts and cell phones. Scientists examined four women between the ages of twenty-one and thirty-nine (which is below the typical age range of women who develop breast cancer). All the patients regularly carried their smartphones in their bras, directly against their breasts, for up to ten hours a day over the course of several years. Despite having no known increased genetic risk factors, each of the women developed tumors in areas of their breasts immediately underlying the phones.

From looking at this data, these researchers agreed that cellular devices appeared to pose a real risk of carcinogenesis. However, the study is considered very limited. For instance, the sample size is a population of only four. Also, no data is available on the number of women who place their cell phones in contact with their breasts and do not develop breast cancer.[67]

In addition to the 2013 University of California study, a 2014 study found that ELF radiation may increase the risk of breast cancer for women before they reach menopause.[68]

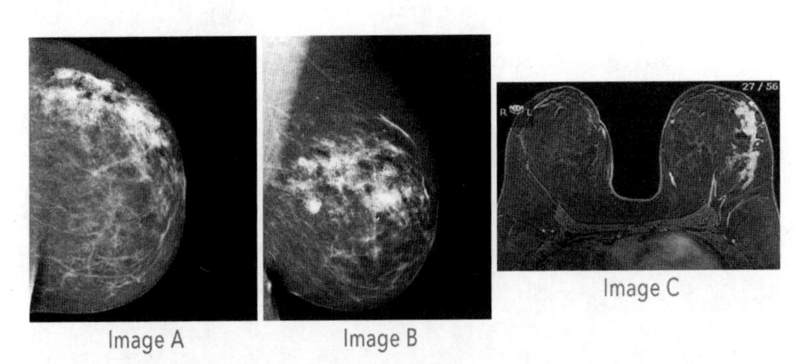

Representative imaging of patient in Case 1. Image A is a mammogram showing clustered calcification corresponding to multiple sites of disease in craniocaudal. Image B shows mediolateral-oblique projections. Image C is an MRI showing extensive nonmass enhancement in the lateral hemisphere of the left breast in segmental distribution.
Photos: West et al., 2013.[69]

Leukemia (Childhood Cancer)

Over the past few decades, dozens of studies have looked at the possible link between EMF exposure and leukemia.[70] Although some studies found no correlation, overall the research shows a slightly increased risk.[71] Homes with high levels of EMF radiation or those near overhead power lines had an increased incidence of children who developed leukemia.[72]

Electromagnetic Hypersensitivity (EHS)

A growing segment of the population reports negative physical symptoms when exposed to EMF radiation emitted from electronic devices, mobile towers, Wi-Fi, and other EMF-emitting sources. This condition is called electromagnetic hypersensitivity syndrome (EHS). Typical symptoms associated with EHS include headaches, fatigue, eye problems, dizziness, stress, sleep trouble, skin issues, gastrointestinal complications, and muscle aches or pains, as well as many others.[73]

This condition is poorly understood and often misdiagnosed, mostly because many of the common symptoms mimic other conditions and the source is hard to pinpoint. We go into greater depth regarding EHS in chapter 5.

Fertility/Reproductive Health

Along with cancer, fertility and reproductivity health issues related to EMF exposure have been one of the biggest research topics over the years. Following are a few areas of study.

Erectile Dysfunction

In 2013, researchers found links between cell phone use and erectile dysfunction. Subsequent large-scale studies have confirmed the initial data and have compelled scientists to recommend exploring the mechanisms involved in this phenomenon.[74]

Pregnancy

A developing organism is sensitive to EMF exposure and vulnerable to the development of radiation-induced effects.[75] The following recent studies illustrate the health effects associated EMF exposure during pregnancy.

In 2012, a Stanford study found that children who were exposed to high levels of EMF radiation while in their mothers' wombs had a 69 percent higher risk of suffering from weight problems and obesity during childhood than children who had low in utero exposure to EMF. The researchers observed that children who were exposed to higher doses of in utero EMF not only had a higher risk of obesity but that this risk kept increasing as the dose of EMF levels increased.

According to the researchers, this is the first epidemiological study that clearly shows a link between increasing EMF exposure, particularly fetal exposure in the mother's womb, to the rapid increase in childhood obesity.[76]

A 2013 study found that prolonged EMF radiation exposure use was associated with a six-fold increase in the risk of miscarriage.[77]

A Chinese study published by the *Journal of Radiation Research* in 2014 found that after being exposed to EMF emissions in the womb, mice showed less movement, indicating increased anxiety. The male offspring also showed decreased learning skills and memory capacity. This study was the first to publish evidence that EMF emissions could induce different effects depending on the gender.[78]

In a 2015 study, researchers found that exposure to a 2.45 GHz frequency of EMF radiation, particularly in the prenatal period, resulted in postnatal growth restriction and delayed puberty.[79] Two other studies that year found that the use of mobile phones while pregnant can be related to early spontaneous abortion.[80]

The fact that this many issues have been found is disconcerting and caution is strongly advised for pregnant women to avoid unnecessary EMF radiation exposure.

Sperm Defects

In 2009, research supported by the Cleveland Clinic's American Center for Reproductive Medicine showed that RF radiation emitted from cell phones may lead to oxidative stress in human semen, causing a significant decrease in sperm motility and viability. The researched speculated that keeping a cell phone in pant pockets can negatively affect spermatozoa and impair male fertility.[81]

A 2011 Argentinian study found that the use of Wi-Fi–enabled laptops significantly decreased sperm motility and increased DNA fragmentation, even from just four hours of exposure. Testing was conducted over a four-hour period and found that 25 percent of male sperm was no longer active, compared to 14 percent from sperm samples stored away from the computer at the same temperature. The results also found that 9 percent of the sperm showed DNA damage, three times the damage experienced by the comparison samples.[82]

In 2013, an American meta-analysis of ten previous studies funded by the Natural Environment Research Council found that mobile phone radiation exposure was associated with reduced sperm motility and viability.[83]

In a study released in 2015, researchers based in Iran studied the effect of short- and long-term exposure of Wi-Fi radiation on male fertility. They found that sperm concentration, motility, and morphology were significantly affected by exposure to Wi-Fi radiation, and the degree of these effects was dependent on length of exposure.[84]

Other studies conducted on rat models have shown that EMF emissions can cause oxidative stress and negatively impact sperm quality by creating malformations and decreasing sperm motility.[85] As a result, the medical community suggests that men avoid putting EMF-emitting technology close to their reproductive organs.

Testes Deformation

EMF radiation has been shown to effect testicular function and structure in male rats. A 2015 study found that there were changes in reproductive organs after one year of exposure to Wi-Fi signals. The researchers concluded, "We suggest Wi-Fi users avoid long-term exposure of RF emissions from Wi-Fi equipment."[86]

A separate study that same year showed alterations in adult rats' testicular structure and biochemistry after being exposed to 900 MHz frequencies for one hour each day over thirty days.[87]

Another 2015 study found that long-term EMF exposure may lead to structural and functional changes of the male testes.[88]

Insomnia/Sleep Issues

In 2008, a joint study involving Wayne State University and researchers in Sweden found that cell phone usage for a period of three hours or more prior to bedtime not only disrupts sleep patterns but may also cause headaches and difficulties in concentration.[89]

A 2010 study found a significant disruption of melatonin in the pineal gland due to EMF exposure. According to the report, "The pineal gland is likely to sense EMFs as light but, as a consequence, may decrease the melatonin production."[90] Melatonin regulates the body's sleep-wake cycle, and if normal production is disrupted, it could possibly lead to long-term health effects.

In 2013, the United States government collaborated with the Egypt Foreign Ministry to fund research on the effect of cell phone radiation on the central nervous system. The study found that exposure to EMF radiation for one hour a day for one continuous month caused rats to experience a delay or latency period before they experienced rapid eye movement deep sleep, which is necessary for restful sleep.[91]

More recently, a 2015 study found that exposure of rats to 1.8 GHz frequencies can affect the circadian rhythm while decreasing the daily production of melatonin. It also showed that superoxide dismutase and glutathione peroxidase (which help prevent damage to cells) were decreased.[92]

Liver Damage

A Chinese study published in May 2014 found that 900 MHz cell phone emissions could cause liver damage in rats by affecting the expression of the Nrf2 (a protein that regulates the expression of antioxidant proteins) and inducing oxidative injury.[93]

Tinnitus

In a review of over 165 clinically relevant studies, a Brazilian study found that cell phone radiation emissions may trigger the onset or worsening of tinnitus.[94]

Toasted Skin Syndrome (Erythema ab igne)

Several studies have found that heat and radiation exposure from laptops can lead to erythema ab igne, otherwise known as toasted skin syndrome, a permanent reddish-brown hyperpigmentation on the surface of the skin.[95]

Toxic Exposure

There are some theorists that believe EMF exposure may be akin to an environmental pollutant. Toxic effects are purported to be exacerbated by other atmospheric toxins, such as chemicals. Exposure to both may lead to free radical formation and oxidative stress with corresponding health consequences. Some research suggests that exposure to both EMF radiation and chemical toxins combine to create a negative impact greater than the sum of its parts.[96]

Research for Yourself

As you can see, a significant amount of research already exists indicating that EMF exposure causes negative biological changes, and our understanding increases more and more each day. While the full implications of long-term EMF exposure are still being studied, it is a good idea to stay abreast of the latest information. We encourage you to research publicly accessible information and to take necessary precautions. You can easily search online for the latest health studies on websites such as PubMed, Science Direct, or Google Scholar. Although we feel the evidence speaks for itself, safety is relative. How safe you feel should be dictated by your own personal guidance. It should be an informed decision that you make when you are equipped with all the data.

Chapter 5

SPECIAL CASES: CHILDREN AND THE ELECTROMAGNETIC HYPERSENSITIVE

"Children are more vulnerable to RF/MW radiation because of the susceptibility of their developing nervous systems. RF/MW penetration is greater relative to head size in children, who have a greater absorption of RF/MW energy in the tissues of the head at WI-FI frequencies. Such greater absorption results because children's skulls are thinner, their brains smaller, and their brain tissue is more conductive than those of adults, and since it has a higher water content and concentrations."

–Dr. David O. Carpenter, MD, director of the Institute for Health and the Environment at the University at Albany and coeditor of the *BioInitiative Report*

Children

Our children are using electronics at younger and younger ages.
Photo: © 123rf.com/wavebreakmediamicro.

In the United States, "family plan" deals make it easy for parents to afford cell phones for their children. Cell phones and tablets are advertised as a way to bring families closer together and help them stay connected. While this may be true to some extent, it means that children are now using these electronic devices from an extremely young age and over longer periods. In 2015, a survey commissioned by the American Speech-Language-Hearing Association reported that 59 percent of children age eight or younger use tablet computers, 52 percent use smartphones, and 48 percent use laptops at home.[97]

In science and medicine, it is widely accepted that kids are not just "miniature adults." Children are different on physical, chemical, and biological levels, because their bodies are still developing. According to the *Stewart Report*, published in 2000 by the Independent Expert Group on Mobile Phones, children are particularly at risk from EMF exposure because a child's body absorbs up to 60 percent more energy per pound of body weight than an adult's. A one-year-old's body can absorb around double the amount.[98] This is most likely due to the higher water content in children's tissues than adults'.[99] Their hormones, enzymes, and cellular responses to environmental toxins are all different from adults', as well. Children are more fragile, and stressors (such as exposure to radiation from electronics) can have profound effects on their developing bodies. These effects can be cumulative, and will span the child's lifetime at rates never seen by today's adults or by previous generations.

Recent Research Related to Children

Unfortunately, until now, EMF research has been focused on adults in a clinical environment. However, EMF research on children should be a primary concern because they are the ones that are exposed during critical developmental periods of their lives. Although data is limited, recent research findings show the harm EMF radiation can have on children.

In Australia, for instance, several studies investigated the effects of cell phone radiation exposure on cognitive function in adults. In 2010, this led to the investigation of children and adolescents, with special interest focused on their developing nervous systems. Over two hundred students were assessed, once at the beginning of the year and again at the end. The results showed some changes in cognitive function associated with increased cell phone RF radiation exposure. Unfortunately, these changes were dismissed as statistically insignificant because the test population was too small and the study scope was too limited.[100]

That same year, German researchers studied over three thousand children, randomly selected from four cities in the south of Bavaria. They took interview data on the children's mental health, sociodemographic characteristics, and other points of interest. The researchers then assessed the test subjects for RF and EMF exposure over the course of twenty-four hours. In total, 7 percent of the children and 5 percent of the adolescents showed abnormal mental behavior.[101]

In 2014, the *Journal of Microscopy and Ultrastructure* published an article, "Why Children Absorb More Microwave Radiation Than Adults: The Consequences," which analyzed past studies on the risk of EMF radiation on children. As the title suggests, past studies found that children are more vulnerable to radiation than adults. Their brain tissues are more absorbent, their skulls are thinner, and their overall body size is smaller. This increases their vulnerability to biological risks, such as brain cancer (which may not form until adulthood), and behavioral problems, such as ADHD. The article noted that fetuses exhibited even greater vulnerability than children and concluded that pregnant women should avoid exposure.[102]

A 2015 study conducted by Igor Yakymenko found that radiation damages DNA over a long period and may lead to formation of cancer cells.[103] Yakymenko made the point that children using cell phones are at greater risk than adults. "[Our] data were obtained on adults who used cell phones mostly up to 10 years as adults," Yakymenko said. "The situation can dramatically differ for children who use cells phone [*sic*] in childhood, when their biology [*sic*] much more sensitive to hazardous factors, and will use it over the [*sic*] life."[104]

How are Kids Affected by Cell Phone Radiation?

To put this EMF radiation risk for children into perspective, we know that when an adult uses a cell phone for long periods, EMF emissions can increase the temperature of the head tissue and bone by as much as 2°F.

Additionally, considering today's standards for the maximum signal strength of cell phones, a phone can penetrate an adult's head up to a few inches. This is disturbing, but it's nothing compared what happens to a child. The same signal can pass completely through a young child's head! Take a look at the following illustration.

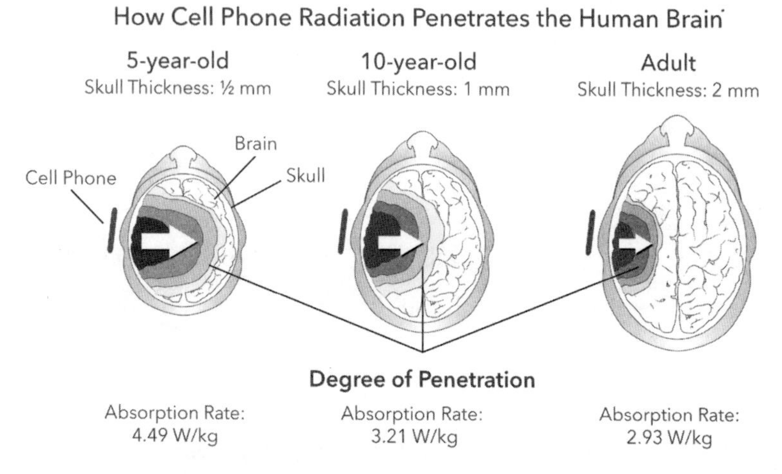

The absorption depths of a 900 MHz GSM cell phone emission in a five-year-old, in a ten-year-old, and in an adult are all very different, with younger brains being the most vulnerable. Data Source: IEEE Transactions on Microwave Theory and Techniques.[105] *Image: R. DeBaun.*

There is limited understanding of this thermal impact from EMF radiation on children. Even more worrisome, though, is that current FCC standards do not take into account the non-thermal, biological impact on children. The biological impact is what should worry us. Short-term exposure can cause redness and burning sensations, but long-term exposure could cause permanent damage, the full extent of which is still unclear. We are just beginning to understand the impact of EMFs on adults, but we can only imagine the impact on children.

Should Wi-Fi be Allowed in Schools?

Many parents and experts are concerned about the use of Wi-Fi in schools. Image: R. DeBaun.

As Wi-Fi has become more pervasive in our society, it has been installed in schools across the globe, giving many parents pause. Numerous scientists, experts, medical professionals, and advocacy groups are becoming vocal about the possible dangers as well. On the surface, using Wi-Fi in the classroom seems like an obvious advantage for making education more modern and convenient. However, it presents risks to our children's health that should be considered.

Wi-Fi devices use radio transceivers that send and receive RF signals. When a student uses a laptop or tablet in the classroom, the device connects to the Internet via routers or access points and transmits data packets over this signal.

Within the confines of the room, there can be numerous signals saturating a child's body for up to eight hours a day, every day, over many years.

Dr. David O. Carpenter, MD, director of the Institute for Health and the Environment at the University at Albany and coeditor of the *BioInitiative Report*, explains that children can be impacted at EMF exposure levels lower than FCC guidelines (which are based on the height, weight, and stature of a six-foot-tall man).

In a 2012 lawsuit, *Morrison v. Portland Public Schools*, which sought to force Portland Public Schools to remove Wi-Fi, Carpenter wrote in a testimony, "In the context of school development, Wi-Fi exposes building occupants including children and adults constantly from both computers and infrastructure antennas. Duration may be an even more potent contributing factor to RF/MW radiation bioeffects than exposure levels. Chronic, such as all-day, school exposure, is more likely than short and intermittent exposure, such as cell phone use, to produce harmful health effects, and is likely to do so at lower exposure levels."[106]

More recently, in 2014, parents of a twelve-year-old boy filed a lawsuit against Fay School, a Massachusetts boarding school where the boy attended, claiming the school's Wi-Fi network was making their son ill. The parents said their son began frequently experiencing headaches, nosebleeds, nausea, and other symptoms while sitting in class after the school installed a stronger wireless network. Dr. Jeanne Hubbuch, a physician, diagnosed the boy with EHS and wrote in a letter to Fay School that there was no other medical explanation for his symptoms.

Dr. Hubbuch wrote, "It is known that exposure to Wi-Fi can have cellular effects. The complete extent of these effects on people is still unknown. But it is clear that children and pregnant women are at the highest risk. This is due to the brain tissue being more absorbent, their skulls are thinner and their relative size is small." Dr. Hubbuch continued, "Due to biochemical individuality some people are more susceptible to these effects than others."[107]

In another letter sent to the school, Dr. Martin Blank, PhD, wrote, "I can say with conviction, in light of the science, and in particular in light of the cellular and DNA science, which has been my focus at Columbia University for several decades, putting radiating antennas in schools (and in close proximity to developing children) is an uninformed choice. Assurances that the antennas are within 'FCC guidelines' is meaningless today, given that it is now widely understood that the methodology used to assess exposure levels only accounts for one type of risk from antennas, the thermal effect from the power, not the other known risks, such as non-thermal frequencies, pulsing, signal characteristics, etc."[108]

Using technology in schools is important; yet making sure it is safely implemented is crucial. There may be no magic bullet, but thinking twice before using Wi-Fi in schools can be an important step in protecting the futures of children.

As an alternative to Wi-Fi, schools can use wired networks to deliver much safer Internet access in the classroom. Parents can do the same for themselves and their children at home.

Are Tablets in Schools a Good Idea?

It used to be that desktop and laptop computers were the big thing in schools, but now tablets are on the rise. Tablets are easy to use and come with a plethora of free educational apps. They're small, light, and wireless, so they're very portable. They allow students to access the Internet, with its unlimited online resources, at the tap of a finger. Students have an immense library of online information literally at their fingertips.

There are many creative ways educators are using tablets in schools to enhance the learning experience for their students. For example, students may be encouraged to take images of backyard wildlife using a tablet or cell phone. When they return to class, they upload these photos to online citizen-science sites, such as Project Noah. This provides an online platform where unidentified specimens are identified by biologists while at the same time contributing to an online database of species diversity and distribution. Tablets clearly have excellent educational benefits—but what is the cost?

The problem with students using tablets in class is that it contributes to even higher levels of EMF radiation exposure. These devices are small by design and feature touch-screen controls, yet they can usually only be connected to the Internet via a Wi-Fi signal. This means that tablets cannot use a wired connection, so there is no way to reduce EMF exposure when using the Internet. Moreover, as we have already said, children are particularly vulnerable to these risks because of their developing bodies. Many popular tablet manuals actually say to limit the time spent using tablets and to use them at a distance away from the body.[109]

So, the only way for children to completely avoid EMF exposure is to keep the tablets away from their bodies and to place them in airplane mode. However, this is perhaps unrealistic and unreliable. It is a good idea to make sure that your children, as well as their teachers, are aware of the risks of EMFs and that they take the necessary precautions to reduce the risks.

Precautions for Parents

Regardless of the dangers of EMF radiation, it may not be a great idea to let very young children use electronic devices at all. For instance, the American Academy of Pediatrics discourages the use of entertainment media, including televisions, computers, cell phones, and tablets, by children under age two.[110] Other research published by Jenny Radesky, clinical instructor in developmental-behavioral pediatrics at Boston University School of Medicine, indicates that children under the age of three learn better by interacting with other humans than by using mobile devices and videos.[111]

Parents often use these devices to distract or pacify their children, not realizing that these items could be detrimental to their child's development of math and reasoning skills, social skills, and the ability to self-soothe. There are advantages to educational programming such as *Sesame Street*, but only for children much closer to school age.

Select governments and the World Health Organization recognize that EMF research for children should be given priority because of the lack of information in this critical area. In the meantime, multiple countries (including Belgium, France, and India) have already passed laws or issued warnings about the exposure of children to wireless devices. Researchers suggest taking precautionary measures to limit EMF radiation exposure to our young. The broader scientific community needs to invest more time, money, and research to establish a broader acceptance of these dangers.

We already know that research is beginning to show long-term health effects of EMF radiation on children. What will it take to establish a common understanding in both the governmental and the scientific communities that EMF exposure is harmful to children? It's not too late for parents to minimize their children's exposure to EMF. Why take the risk with your child's health and development?

Electromagnetic Hypersensitivity (EHS)

"Sensitivity to Electromagnetic Radiation is the emerging health problem of the 21st century. It is imperative health practitioners, governments, schools and parents learn more about it. The human health stakes are significant."

–Dr. William Rea, MD, founder of Environmental Health Center

"There must be an end to the pervasive nonchalance, indifference and lack of heartfelt respect for the plight of these persons. It is clear something serious has happened and is happening."

–Olle Johansson, PhD, associate professor, Experimental Dermatology Unit, Department of Neuroscience, Karolinska Institute, Stockholm, Sweden

A growing part of the population has reported adverse physical reactions related to EMF exposure. This condition is called Electromagnetic Hypersensitivity Syndrome. Photo: © depositphotos.com/SIphotography.

Electromagnetic Hypersensitivity Syndrome (EHS) is characterized by a variety of nonspecific symptoms triggered by exposure to EMF radiation. Despite long-term recognition by the World Health Organization as a genuine ailment, EHS is still poorly understood and somewhat controversial, because its symptoms can differ from one person to another and can vary widely in severity.[112] Furthermore, although it is widely agreed that people claiming to be electromagnetic hypersensitive experience real symptoms, there is scarce evidence of a causal link between their symptoms and EMF exposure.[113] Some critics even claim that EHS is psychosomatic and is not a real condition.[114]

Despite this, the World Health Organization acknowledges that a significant number of individuals have reported a variety of health problems that relate to EMF exposure. Due to the number of those who experience it, many doctors and scientists have started to research it as well.

A wide range of estimates exist for how prevalent EHS is in the general population. According to Dr. Olle Johansson, PhD, of the Karolinska Institute, "Electrical hypersensitivity is reported by individuals in the United States, Sweden, Switzerland, Germany, Denmark and many other countries of the world. Estimates range from 3% to perhaps 10% of populations, and appears to be a growing condition of ill-health leading to lost work and productivity."[115]

In a Swiss survey conducted in 2004, over 90 percent of respondents reported an average of 2.7 common symptoms related to EMF exposure. Seventy-four percent of respondents identified mobile phone towers as the source of their issues, followed by mobile phones (36 percent), cordless phones (29 percent), and power lines (27 percent). Some individuals reported mild symptoms and reacted by avoiding EMFs in general. Others reported they were so severely affected that they ceased work and had to change their entire lifestyles.[116]

EHS Symptoms

Generally, EHS symptoms will be triggered during exposure to low levels of EMF radiation (for example, feeling a tingling or burning sensation when using a cell phone). If you experience any discomfort upon prolonged use of radiation-emitting devices, it is possible that you are experiencing EHS. The following illustration lists commonly reported EHS symptoms.

Electromagnetic Hypersensitivity Symptoms

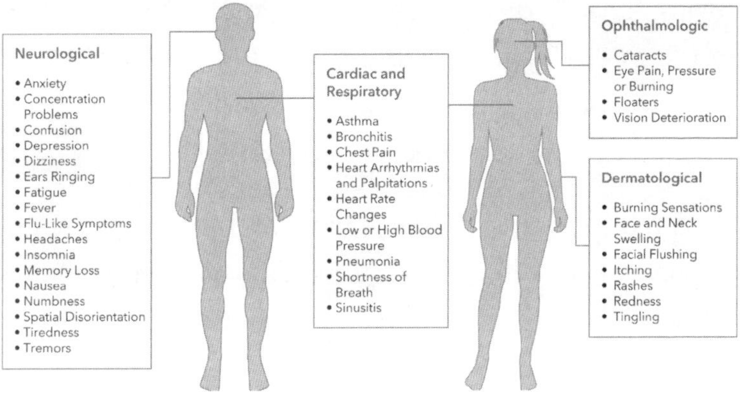

Neurological

- Anxiety
- Concentration Problems
- Confusion
- Depression
- Dizziness
- Ears Ringing
- Fatigue
- Fever
- Flu-Like Symptoms
- Headaches
- Insomnia
- Memory Loss
- Nausea
- Numbness
- Spatial Disorientation
- Tiredness
- Tremors

Cardiac and Respiratory

- Asthma
- Bronchitis
- Chest Pain
- Heart Arrhythmias and Palpitations
- Heart Rate Changes
- Low or High Blood Pressure
- Pneumonia
- Shortness of Breath
- Sinusitis

Ophthalmologic

- Cataracts
- Eye Pain, Pressure or Burning
- Floaters
- Vision Deterioration

Dermatological

- Burning Sensations
- Face and Neck Swelling
- Facial Flushing
- Itching
- Rashes
- Redness
- Tingling

Common symptoms reported by EHS sufferers include neurological, cardiac and respiratory, ophthalmologic, and dermatological issues. Data Source: "Living with Electro Hypersensitivity: A Survival Guide." [117] *Image: R. DeBaun.*

The Evolution of EHS

The term "electrical hypersensitivity" was first used in 1989. However, use of the term "electromagnetic hypersensitivity" began in 1994 to reflect sufferers' sensitivity to magnetic fields, in addition to electric fields. EHS symptoms were observed in people working with radio and electricity as far back as the early 1930s and with military radar in the 1940s.[118]

If reports of EHS have been around so long, and EHS now seems to affect a sizable portion of society, you might wonder why you haven't heard more about it until now. A major reason is that EHS was not really studied until the 1970s, when the general population started using computers.[119] As time has passed, we have begun using electronic devices closer and closer to the body, and because of the preponderance of these devices available today, radiation exposure is becoming more and more common. We now encounter EMFs emitted from our laptops, cell phones, and tablets, as well as from Wi-Fi and Bluetooth, on a daily basis.

EHS has now been recognized as a functional disability in both Sweden and Canada, and if you pay attention to the news, you will notice it is beginning to be recognized by the courts.

For instance, Marine Richard of France won a disability payout from the government due to Wi-Fi sensitivity in August 2015. She stood to claim almost $900 USD per month for three years as a result of her lawsuit. Unfortunately for Richard, she was forced to move into a barn without electricity in a remote region of France to escape EMF radiation.[120]

Studies on EHS are still being conducted, and researchers have called for more investigation to better understand the disorder. To that end, the International EMF Project has been sponsored by the World Health Organization to focus on identifying research needs. The organization also facilitates a worldwide program of studies related to the potential health risks associated with EMF exposure.[121]

The Link to Multiple Chemical Sensitivity (MCS)

A case has been building that EMF radiation exposure is actually an environmental pollutant. Interestingly, EHS resembles multiple chemical sensitivity (MCS), a disorder associated with low-level environmental exposure to chemicals. If someone is electrically sensitive, they are also likely to be multiple chemical sensitive and vice versa.

According to medical science writer Helke Ferrie, "The symptoms of electropollution-induced sickness involve all organs with many debilitating symptoms, from skin rashes to cancer; they are part of the Multiple Chemical Sensitivity (MCS) spectrum."[122]

The two are commonly characterized by a range of nonspecific symptoms that lack a toxicological or physiological basis. In both cases, it is likely that the body's cells shut down and stop communicating with each other. In the case of MCS, this could be caused by volatile organic compounds. In the case of EHS, the culprit could be a cell phone signal that causes headaches, tingling, and so forth.

What You Can Do if you are Electrically Hypersensitive

EHS is a real concern, and its harm should not be underestimated and those who suffer from it should not be shrugged off. In fact, it can be a disabling problem for affected individuals. If you think you or your loved ones are experiencing EHS, you should take necessary precautions to ensure it does not affect your quality of life. Unfortunately, there is currently no cure to rid you of this condition. We live in a time where EMF radiation is around us all the time.

The good news is you can start to limit any unnecessary EMF exposure right away through a variety of means. Following is a list of tips that EHS sufferers can use to protect themselves and to live more comfortable, healthier lives.

Electromagnetic Hypersensitivity Tips

1. Keep a safe distance and limit exposure to electronics, appliances, and electrical equipment, such as computers, microwave ovens, refrigerators, induction stoves, TVs, hair dryers, and so on. To go further, try unplugging them when not in use.
2. Use a wired network instead of Wi-Fi to connect to the Internet. Wired networks (using Ethernet cables) are not only safer but also faster.
3. Keep your bedroom or sleeping environment free of all wireless and EMF-emitting devices or electronics.
4. For phone calls, use a landline instead of a cell phone.
5. Remove fluorescent lighting and replace it with incandescent lighting.
6. Utilize EMF-blocking devices, fabrics, and paints that are capable of blocking both ELF and RF radiation.
7. Consider contacting a local EMF expert who can advise you on removing EMF-emitting sources from your home.
8. Consult your doctor, naturopathic practitioner, or other health professional on possible EHS exacerbators, nutritional improvements, and lifestyle changes that can help lessen symptoms.
9. If possible, do not live near power lines, cell phone towers, electrical substations, and other similar buildings. This is more difficult in dense city areas than rural ones, but it is more ideal for EHS sufferers.

Chapter 6

EMF SAFETY STANDARDS

"Radio frequency radiation and other forms of electromagnetic pollution are harmful at orders of magnitude well below existing guidelines. Science is one of the tools society uses to decide health policy. In the case of telecommunications equipment, such as cell phones, wireless networks, cell phone antennas, PDAs, and portable phones, the science is being ignored. Current guidelines urgently need to be re-examined by government and reduced to reflect the state of the science. There is an emerging public health crisis at hand and time is of the essence."

–Dr. Magda Havas, PhD, associate professor, environment and resource studies, Trent University

In the United States and other countries across the world, safety standards for non-ionizing EMF radiation are set by government agencies. The scientific evidence is beginning to tell us that these safety standards may be insufficient and should be reconsidered. Before we get into EMF safety standards, we want to talk a little bit about the difference between analog and digital signals.

Analog vs. Digital Signals

Over the past thirty years, technology has slowly transitioned from analog signal processing to digital signal processing.

Analog Signal

An analog signal is represented as a continuous smooth wave with an infinite number of possible values. An example is illustrated below.

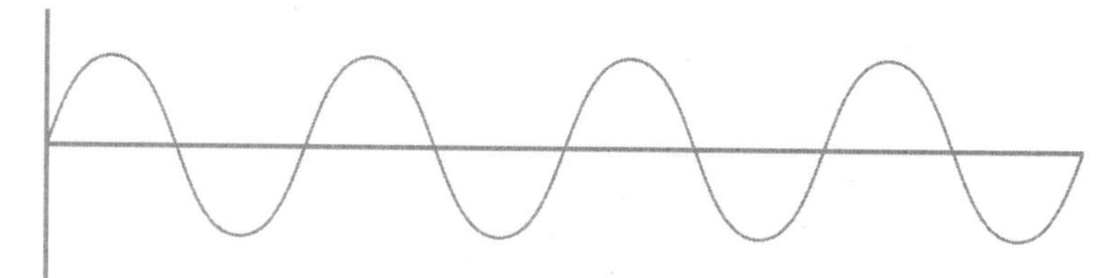

Image: R. DeBaun.

Back in the 1970s, when cell phones were first invented, most electronics used analog signal processing to communicate. Even after mobile phones hit the consumer marketplace in the 1980s, they transmitted via analog signals. It was not until 2G (or second generation) digital networks began to emerge in the 1990s that cell phones started transitioning to digital signal processing. Today, all cell phones transmit digitally.

Digital Signal

Unlike analog signals, digital signals are transmitted in a discontinuous on/off pattern, with a limited set of possible distinct values of information. The image below shows what a digital signal looks like.

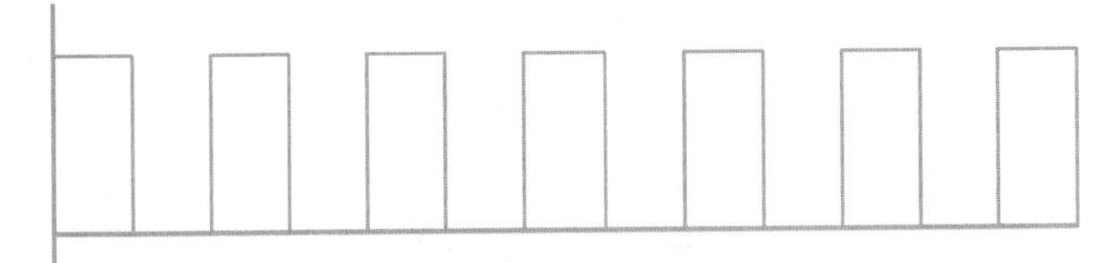

Image: R. DeBaun.

Digital signal processing was first utilized in the 1960s and 1970s when digital computers became available, but it was limited to radar and sonar, oil exploration, medical imaging, and space exploration.[123] It wasn't until 1978 that digital signal processing would make its way into consumer electronics, and it came in the unlikely form of a children's educational toy you might remember fondly—the Speak & Spell. Developed in the late 1970s by Texas Instruments, the Speak & Spell utilized the industry's first digital signal integrated processor—the TMS5100—to generate speech audio from text typed into the toy by a child.[124] Basically, it was the first speech synthesizer.

This photo shows E. T. holding a Speak & Spell he used to "phone home." Photo: Universal/The Kobal Collection.

And not only did the Speak & Spell go on to help teach millions of children how to spell, but it also helped E. T. phone home![125] The Speak & Spell marked the first milestone in the use of digital signal processing that we see in widespread consumer, industrial, and military applications today.

The main advantage of digital signals over analog signals is that the precise signal level of the digital signal is not critical. Digital signals are immune to imperfections of electrical systems, which can disrupt analog signals. For example, digital compact discs are more robust than analog vinyl records. Digital signals can also be turned into codes, which are more secure, and can also be processed by cheap digital circuit components. Digital signals use less bandwidth, and more information can be stored in less space. They enable transmission signals over long distances, and they allow simultaneous multi-direction transmission.

So, what is the main reason we bring up the difference between analog and digital signals?

If you think of an analog signal as a twenty-pound metal rod and you stick it on a concrete sidewalk, the concrete will not break. However, if you take that same metal rod and consistently tap it into the sidewalk like a digital signal's on/off pattern, the concrete will eventually fracture. Another analogy would be placing slight pressure on glass: If you press into glass, nothing will happen. But if you repeatedly hit it, the glass will eventually crack and ultimately break. This is what digital signals are doing to your body. They are continually striking your body's cells, which finally break down over time.

Technology has transformed significantly. You would think that standards for EMF safety would have been revised as technology has evolved. Unfortunately, this has not been the case.

Specific Absorption Rate (SAR)?

In the United States, the FCC regulates safety exposure limits for RF radiation exposure with a measurement called specific absorption rate (SAR). SAR is the amount of RF radiation that the human body (e.g., the head) absorbs when exposed to an electronic device. The SAR value is established by measuring the short-term thermal effect on the area of the body closest to the radiating source and is defined as the amount absorbed per unit of tissue in watts per kilogram (W/kg).[126]

$$SAR = (\sigma E^2 / 2\rho)$$

SAR = Specific absorption rate (W/kg)

E = Peak amplitude of the electrical field in human tissue (V/m)

σ = Tissue conductivity (siemens per meter [S/m])

ρ = Tissue density (kilogram per cubic meter [kg/m^3])

The SAR value will vary heavily depending on which part of the body is exposed to RF radiation as well as the location and geometry of the source. Tests must be made with each specific source in the intended position of use, such as a mobile phone against the head.

While people mostly associate SAR values with cell phones, governmental agencies actually require manufacturers to test SAR levels of any electronic device with wireless transmission capabilities. In the United States and Canada, the SAR limit for mobile devices used by the public is 1.6 W/kg averaged over 1 gram of tissue for the head.[127]

In Europe, the publicly funded International Commission on Non-Ionizing Radiation Protection (ICNIRP) creates SAR standards. In 1998, the ICNIRP established the limit of 2 W/kg averaged over 10 grams of head tissue.[128] In Australia, the Australian Radiation Protection and Nuclear Safety Agency (ARPANSA) sets the SAR limit at 1.6 W/kg averaged over 10 grams of head tissue.[129]

Note that the standards in Europe and Australia are safer because the energy is spread over an area ten times larger than the standard used in the United States.

Recommended SAR Levels

Region/Country	Guideline	Limit in W/kg
United States	FCC (1996)	1.6 W/kg in 1 g of tissue
Europe	ICNIRP (1998)	2.0 W/kg in 10 g of tissue
Australia	ARPANSA (2002)	2.0 W/kg in 10 g of tissue

Why SAR Isn't Adequate

SAR is considered to be an insufficient standard for many reasons. Despite vast changes in technology, SAR standards have not been updated since 1996 in the United States. To make matters worse, the standards were developed based on animal studies from the 1980s. SAR was also established solely to provide guidance for minimizing thermal effects of RF emissions on the body, not the nonthermal biological effects.[130] Finally, SAR does not effectively take into account the effects of exposure to ELF radiation.

Thermal Impact

SAR is used to gauge the short-term thermal effects of RF radiation by measuring temperature increases in the part of the body exposed. For instance, when holding a cell phone to your head for a long period, the immediate area of the head that it touches can heat up. This thermal impact to the head's cells is similar to a reaction that meat has when cooked in a microwave oven. That's because a microwave oven operates in the same range as the signals emitted by a cell phone, just not as strong.

A 2015 study found that radiation emissions can heat up a head's tissue by as much as 1.36°F after only fifteen minutes of using a Samsung Galaxy Note 3.[131] In the short term, exposure to these emissions can cause redness and burning sensations, yet in the long term, it can cause more lasting damage.

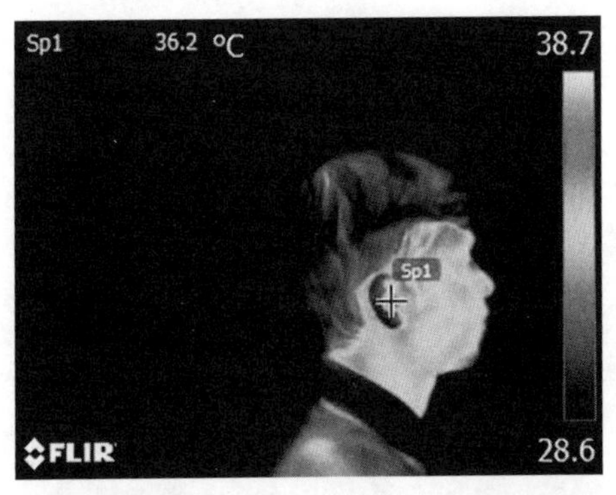

This image shows a human head before exposure to mobile phone radiation.

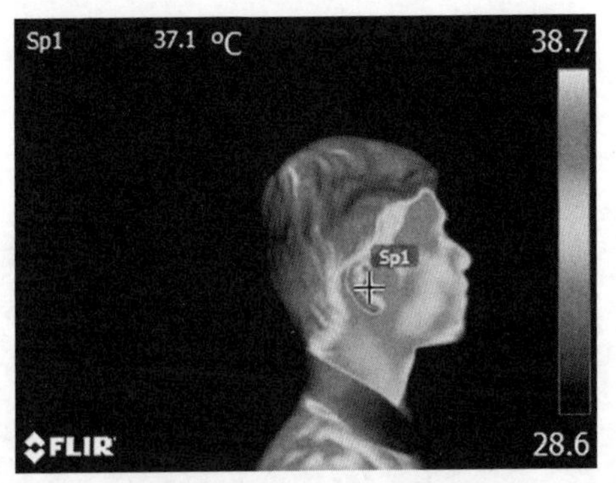

This image shows thermal impact on a human head after exposure to mobile phone radiation for fifteen minutes.

Photos: Varshini Karthik et al., 2015.[132]

Biological Impact

Perhaps an even greater concern than the thermal impact of EMF radiation is the nonthermal, biological impact that it can have.[133] This is what SAR is not good at measuring. Biological impact is not necessarily sensed by the body in the same way thermal exposure is. There may be no burning sensation that can be felt right away. However, biological impacts can have both short- and long-term implications. Over time, cells will mutate under constant EMF radiation exposure. When cells are impacted, they can close themselves off, isolating themselves from neighboring nourishing cells. Cells can also change in other unexpected ways, which create dangerous unpredictability. Science is only beginning to identify the mechanism of cellular changes, but these types of changes have already been linked to cancer, neurological disease, and numerous other health risks.

Other SAR Inadequacies

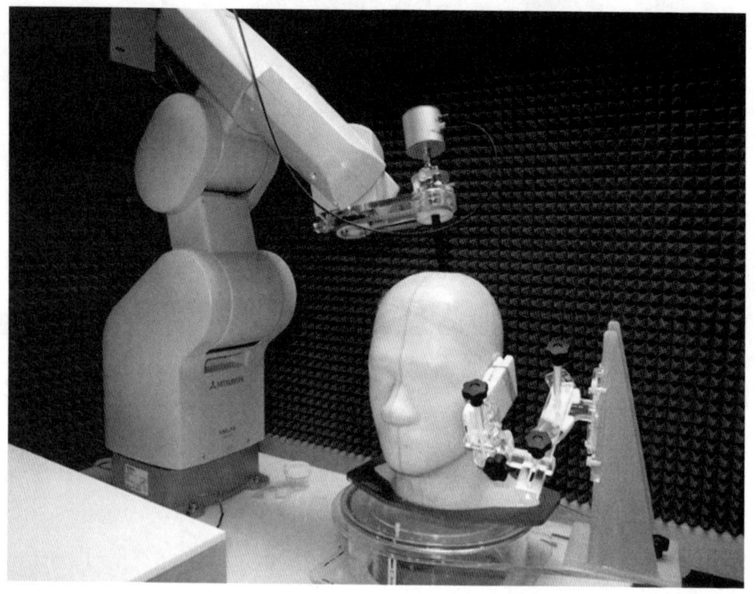

These tools sit in an SAR testing lab. Photo: Wired/Priya Ganapati.[134]

Technology has changed enormously since SAR was first established. Not only does SAR fail to measure the non-thermal biological impact of EMF radiation to the body, it does not account for the ELF radiation emitted by modern smartphones. If that is not concerning enough, SAR tests are not designed with every citizen in mind.

To test SAR, cell phone manufacturers simulate the amount of cell phone radiation absorbed using a plastic model of a head called a specific anthropomorphic mannequin (SAM), a representation of the top 10 percent of US military recruits in 1989.[135] Since this model head is based upon a six-foot-two man who weighs 220 pounds, it only represents a small fraction of the population. In fact, SAM is larger than 97 percent of all cell phone users![136] It does not represent those who are especially susceptible to the damaging effects of EMF exposure, such as children, pregnant women, or senior citizens.

As Om P. Gandhi, professor of electrical and computer engineering at the University of Utah, puts it, "SAM uses a fluid having the average electrical properties of the head that cannot indicate differential absorption of specific brain tissue, nor absorption in children or smaller adults. The SAR for a 10-year old [*sic*] is up to 153% higher than the SAR for the SAM model. When electrical properties are considered, a child's head's absorption can be over two times greater, and absorption of the skull's bone marrow can be ten times greater than adults."[137] In other words, testing with the current standard does not reflect typical, real-world cell phone usage by the general populace.

In fact, even the FCC acknowledges that there may be reason for concern. People who do not feel that adequate protection is provided by SAR are advised to practice precaution. According to the FCC website, "For users who are concerned with the adequacy of this standard or who otherwise wish to further reduce their exposure, the most effective means to reduce exposure are to hold the cell phone away from the head or body and to use a speakerphone or hands-free accessory."[138]

A Call for New Standards – *The BioInitiative Report*

A collection of independent scientists and health experts from around the world, the BioInitiative Working Group, is addressing the issues of EMF safety and standards head-on.

In 2007, and again in 2012, the group published the *BioInitiative Report*, a comprehensive assessment of the health risks associated with exposure to both RF and ELF radiation. The report, which documents more than eighteen hundred research studies, calls for updated and stricter EMF safety standards to be put in place. It also highlights that young children, the elderly, and pregnant women are particularly vulnerable to the associated EMF health risks.

The *BioInitiative Report* stresses that current standards are inadequate and need to be urgently updated in order to protect people from the dangers of EMF exposure.

"There is now much more evidence of risks to health affecting billions of people world-wide," says Dr. David O. Carpenter, coeditor of the 2012 report. "The status quo is not acceptable in light of the evidence for harm."[139]

The *BioInitiative Report* is a great step in the right direction. The main strength of the *BioInitiative Report* is that it has been prepared independently of governments, existing regulatory bodies, and industry companies, many of whom have clung to old standards. Precisely because of this, the *BioInitiative Report* presents a balanced, evidence-based public health policy assessment.

We highly suggest heading over to www.bioinitiative.org and reading the entire report yourself.[140]

High-Speed 5G Wireless Rollout

If standards are not adequate now, they will be grossly inadequate when faster wireless networks are deployed, allowing for much stronger transmissions. On July 14, 2016, the

FCC announced the adoption of new rules allowing for the next generation of wireless broadband operations. These regulations make the United States the first country to start the rapid expansion to 5G networks and technologies by opening up what is known as the millimeter wave spectrum, which uses *extremely* high frequencies of 24 GHz and above. More specifically, the rules open up nearly 11 GHz of frequencies, including Microwave Flexible Use service in the 28 GHz (27.5–28.35 GHz), 37 GHz (37.0–38.6 GHz), and 39 GHz (38.6–40.0 GHz) bands, as well as a new unlicensed band at 64.0–71.0 GHz.[141]

In a question-and-answer session following the announcement, environmental health advocate Kevin Mottus asked FCC Commissioner Tom Wheeler, "With the NTP study showing wireless causes cancer sub-thermally, how can you proceed with more wireless expansion, with FCC standards only recognizing thermal effects, ignoring thousands of studies showing cancerous effects, neurological effects, reproductive harm, immune system disorders . . . ?"[142]

Without answering the very pertinent question, Wheeler cut off Mottus and quickly moved on to a question from a Bloomberg reporter. This is troubling, as it seems the FCC is moving full steam ahead into uncharted waters, unconcerned about the possible health dangers that may be exponentially greater than with those of current frequencies. If health standards are inadequate now, they will be nonexistent for 5G.[143]

"To expand wireless at this point, without adequate safety standards, is absolutely irresponsible," Mottus later told a reporter.[144]

The FCC seems unconcerned. At a National Press Club luncheon one month before the 5G announcement, Wheeler stated, "Unlike some countries, we do not believe that we should spend the next couple of years studying what 5G should be or how it should operate and how to allocate spectrum, based on those assumptions." He continued, "We won't wait for the standards to be the first, to be first developed in the sometimes arduous standards setting process or in government led activity."[145]

If Wheeler's vision comes to fruition, not only is 5G worrying on a health level, but at a privacy and ethical level too. Apparently, everything will be connected with microchips when the next phase of broadband is fully implemented. According to Wheeler, "If something can be connected it will be connected in a 5G world . . . with the predictions of hundreds of billions of microchips connected in products from pill bottles to plant waterers, you can be sure of only one thing: the biggest internet of things application has yet to be imagined."[146]

Chapter 7
PUBLIC WARNINGS

"Exposure to radio-frequency, or RF, radiation is a major risk of cellphone use. Manufacturers have a legal duty to provide warnings that are clear and conspicuous when products raise health and safety concerns. But, typically, RF safety instructions are buried in user manuals with tiny print, hidden within smartphones, or made available on the Internet."

–Dr. Joel Moskowitz, PhD, director of the Center for Family and Community Health, University of California, Berkeley

As we all know, many consumer products come with warning labels that explain any health hazards associated with their use. From microwaves to household chemicals, governments have regulated industry and cautioned the public about potential health risks from certain products. Nevertheless, most portable electronic devices do not carry warning labels, even though evidence of potential harm continues to grow. Unless it's required by law, manufacturers and retailers will rarely volunteer overt warnings. Like the tobacco companies before them, the telecom industry in the United States is vehemently opposed to precautionary labeling. Recently, though, some headway has been made.

In May 2015, Berkeley, California, passed the first law in the United States requiring cell phone retailers in the city to place warning notices informing consumers of the possible health risks associated with RF radiation exposure.

The landmark "Right to Know" ordinance required the notice to say, "To assure safety, the Federal Government requires that cell phones meet Radio Frequency (RF) exposure guidelines. If you carry or use your phone in a pants or shirt pocket or tucked into a bra when the phone is ON and connected to a wireless network, you may exceed the federal guidelines for exposure to RF radiation. This potential risk is greater for children. Refer to the instructions in your phone or user manual for information about how to use your phone safely."[147]

Soon after the law was passed, the wireless industry trade lobby, CTIA (formerly Cellular Telephone Industries Association), filed a federal suit in the Northern District of California, claiming the law violated the First Amendment.

However, Berkeley was able to fend off the lawsuit by removing the line "This potential risk is greater for children." In his order, presiding judge Edward Chen wrote, "On the first preliminary injunction factor, the Court cannot say that CTIA has established a strong likelihood of success on the merits with respect to its First Amendment claim. Nor has it raised serious question on the merits. While the sentence in the Berkeley ordinance regarding the potential risk to children is likely preempted, the remainder of the City notice *is factual and uncontroversial and is reasonably related to the City's interest in public health and safety*. Moreover, the disclosure requirement does not impose an undue burden on CTIA or its members' First Amendment rights"[148] (emphasis added).

The "Right to Know" ordinance finally took effect in 2016[149] and is seen by activists as a model for other city governments to follow.

Berkeley's "Right to Know" cell phone radiation warning took effect in 2016. Photo: NBC Bay Area.[150]

Creating Doubt

Certainly, the mobile industry will not take future warning laws lying down. Its stance is that cell phone emissions are completely safe, and it will continue to downplay research showing the health dangers of cell phone radiation. This is one of the topics of the 2001 book *Cell Phones: Invisible Hazard in a Wireless Age*, written by noted epidemiologist Dr. George Carlo, PhD, JD. In his book, Dr. Carlo wrote about his six years (from 1993 to 1999) as the chief scientist for CTIA, where he was hired to oversee studies investigating the safety of cell phones and their possible relationship to brain tumors.

Dr. Carlo reviewed past studies conducted by leading scientists from around the world, and his conclusions were not good news for the wireless industry. He discovered that 2.4 GHz radiation (cellular, Wi-Fi, and so on) appeared to cause generation of micronuclei in human blood samples. This was significant because other studies at the time were showing a strong link between micronuclei in blood and cancer, with children being at particularly high risk.

After presenting his findings, Dr. Carlo asserts that Tom Wheeler, then president of CTIA, intentionally redirected funding away from the project. In fact, with Dr. Carlo's findings in direct conflict with the CTIA, Wheeler publicly discredited Dr. Carlo. He even went on ABC's *20/20*, aggressively arguing that there was no link between the use of wireless phones and health issues.[151] Yet, it is not unreasonable to see why CTIA would not want this information spreading. It is their job to lobby for the interest of telecom companies, and Carlo's findings were extremely damaging.

It is also worth noting that as of 2016, Wheeler (the man who discredited research on the safety risks of cell phones) heads the FCC and is not expected to address the need for updated safety standards.

Parallels to Cigarettes?

In 2004, Dr. Robert N. Proctor, Stanford professor and historian, was asked to give expert testimony in the *Tobacco and Health Report* for the federal court case, *The United States of America v. Philip Morris*. In a window he refers to as "time lag," Dr. Proctor stated that there was clear evidence that cancer and other health issues related to cigarette smoking might take twenty, thirty, or even forty plus years to appear. This explains why there wasn't a massive increase in lung cancer diagnoses in the earlier part of the twentieth century when cigarette smoking first became popular. It took many years of exposure to cigarette carcinogens for symptoms to reveal themselves.[152] Many believe we are now experiencing that same time lag when it comes to EMF health issues.

While the scientific community compiled evidence over decades and decades showing that cigarettes caused cancer, the tobacco companies continued to deny any link. They spent massive amounts of money for a grand-scale PR deception campagn. At the same time, they were offering their own studies that were either misleading or false. Tobacco companies were eventually found responsible for withholding knowledge of the serious health issues caused by cigarettes, yet many decades passed before there was any public acknowledgement. In fact, it was not until 1965 that the US Congress passed the Federal Cigarette Labeling and Advertising Act, which required a health warning to be placed on all cigarette packs.[153] Even after the labels were added, tobacco companies continued to deny any risks.

So, while the dangers of EMF radiation may not be common knowledge currently, this may not be the case in twenty years. With the past serving as a reminder, maybe proactive safety measures should be used early to avoid health issues in the future. Unfortunately, we probably will not see adequate EMF warning labels on mobile devices anytime soon, but it still makes sense to be cautious when using them close to the body.

Product Warning Labels

Although there are currently no overt product warning labels required on mobile device packaging, you can, however, find some cautionary language in the fine print. If you look closely at the package insert of the Apple iPhone 7, for instance, you'll read, "To reduce exposure to RF energy, use a hands-free option, such as the built-in speakerphone, the supplied headphones, or other similar accessories. Carry iPhone at least 5mm away from your body to ensure exposure levels remain at or below the as-tested levels."[154]

This is better than nothing, but how many consumers actually read product inserts? On top of that, the warning refers to the phone being within SAR limits, which we feel is now an outdated and inadequate standard.

Some argue that this kind of language is placed into the fine print to avoid any sort of liability if health issues related to use of the product actually occur. While this may be true, electronic device manufacturers, just like tobacco companies, don't necessarily need to worry about liability. The statistical nature of cancer, and other health issues to which a product may contribute, leave these manufacturers nearly untouchable. Despite overwhelming evidence against them, this is what enabled the tobacco industry to fight off lawsuits for so long. Since we use so many electronic devices throughout our lives (we change phones every year or two, for example), it's nearly impossible to pin down which device caused a long-term health issue.

Chapter 8

HOW TO PROTECT YOURSELF

"There are many examples of the failure to use the precautionary principle in the past, which have resulted in serious and often irreversible damage to health and environments. Appropriate, precautionary and proportionate actions taken now to avoid plausible and potentially serious threats to health from EMF are likely to be seen as prudent and wise from future perspectives."

–Jacqueline McGlade, chief scientist, United Nations Environment Program

We hope that you have learned a lot about EMF radiation so far. You now know that it surrounds us in our everyday life and is an inherent part of the tools we use in today's world. So, if you want to live in modern society instead of like a caveman, it is impossible to get rid of or completely avoid it. Limited EMF radiation exposure is fairly safe; it is overexposure that causes harm.

What can you do right now to protect yourself and your loved ones from the possible health risks of EMF overexposure?

To limit the amount of EMF exposure in our daily lives, we should apply the Precautionary Principle. According to

UNESCO's definition of the Precautionary Principle, "When human activities may lead to morally unacceptable harm that is scientifically plausible but uncertain, actions shall be taken to avoid or diminish that harm."[155]

The Top Three Ways to Minimize EMF Health Risks

Luckily, there are three simple things that you can do to limit your exposure. To minimize the risk of EMF radiation, these are your most highly effective options: (1) create distance, (2) reduce time, and (3) shield yourself.

Create Distance

EMF radiation's strength is measured by its intensity (wavelength) relative its distance away from the source. So, if you want to minimize exposure to EMF radiation, the first thing you should do it create *distance* between you and the source to lessen the intensity.

An electromagnetic field radiates spherically (i.e., goes in all directions), getting weaker as it progresses away from the source. Therefore, radiation's effect on you is stronger the closer you are to an electronic device. This harmful effect decreases as you move away from it. In physics, this is called

Newton's inverse square law. The intensity is proportional to $1/\text{Distance}^2$.

Newton's Inverse Square Law

$$\text{Intensity} \propto \frac{1}{\text{Distance}^2}$$

As a general rule of thumb, if you are one foot away from the source, you experience about one-fourth the amount of radiation exposure as you would if you were touching the source. If you are two feet away from the source, you now experience around one-sixteenth the amount of radiation.

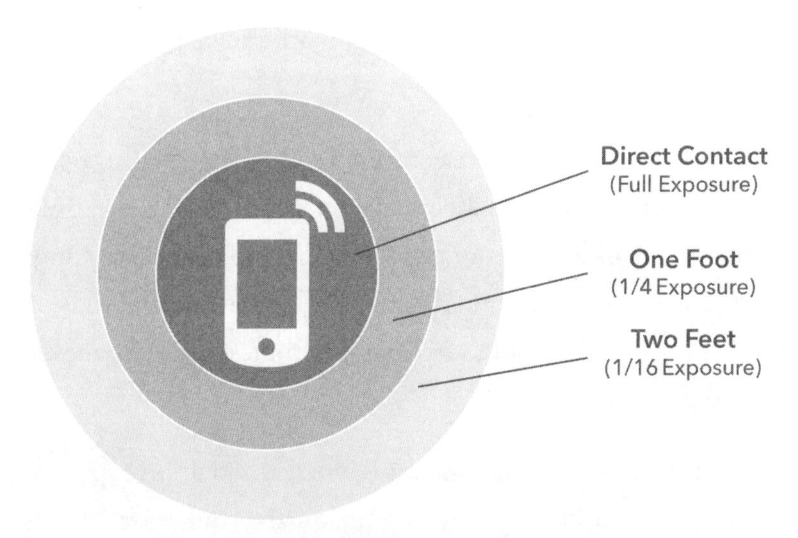

EMF radiation intensity decreases as it moves away from the source. Image: R. DeBaun.

Distance is the greatest weapon you have at your disposal for protecting yourself and your family against EMF radiation. Even a small increase in space between you and your laptop, cell phone, or tablet can make a tremendous difference. All that's needed is a few inches to start minimizing risks!

Reduce Time

The second option you have to decrease your exposure to EMF radiation is *time*. This means reducing the amount of time you spend using radiation-emitting devices.

EMF radiation exposure takes time to have an effect. It's a steady stream of energy that changes cells subtly over time. Exposure for short periods of time are not dangerous, but risks can be cumulative.

If you can help it, use electronic devices close to your body for only brief durations. Begin by limiting the amount of time you spend talking on your cell phone next to your head. Only use your laptop or your tablet on your lap for short periods of time. Take breaks from using your electronic devices whenever possible.

The chart below demonstrates ionizing radiation exposure over time, but also applies to non-ionizing radiation, which works the same way.

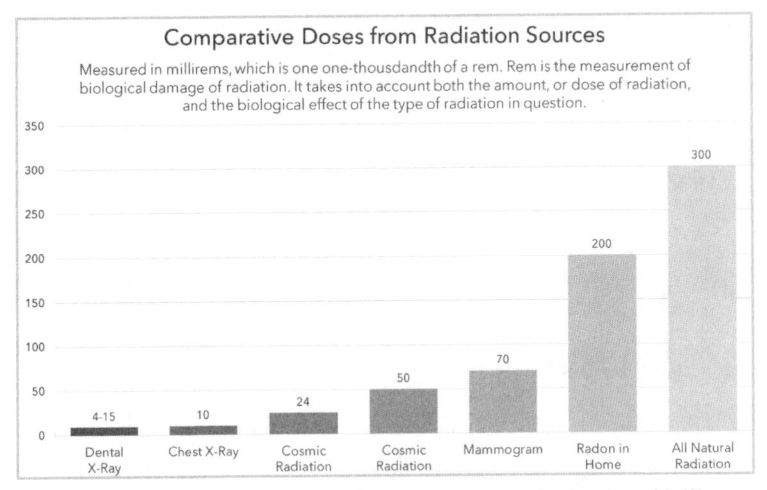

Data Source: National Institute for Occupational Safety and Health.[156]
Image: R. DeBaun.

Shield Yourself

It is possible to guard against EMF emissions from electronic devices by using *EMF shielding*. EMF shields can limit, or completely prevent, radiation from reaching your body if placed between you and the source. A number of commercially available products are on the market. Some are useful, but most still will require vetting. Each product needs to be evaluated on a case-by-case basis. As an informed consumer, you can consider some of these options.

What to Look for in EMF Shielding

When researching EMF shields, at the bare minimum, make sure it will shield you from both ELF and RF radiation.

This includes:

- radiation from device hardware (ELF radiation);
- Wi-Fi radiation (RF radiation);
- cellular radiation (RF radiation);
- Bluetooth radiation (RF radiation); and
- heat radiation.

Check to see if the products have had independent testing performed by third-party labs. While companies may have performed some tests themselves, it is important to get testing by outside labs, preferably from labs which are verified by the FCC. Verified labs have highly accurate testing equipment and environments free from ambient emissions. Outside of a laboratory, it is very difficult to eliminate ambient emissions from other sources. Ambient radiation can highly influence or significantly distort measurements, so unless testing is conducted in an isolated environment, accurate readings are difficult to obtain.

What You Should Question

With the increased awareness of EMF health risks, there has been an influx of "solutions" into the marketplace that promise protection. Unfortunately, some of these products don't do much of anything at all; they often lack solid science to back up their claims or they can't prove it. Sometimes it seems they don't even realize that their products don't do what they claim.

Not all companies are *intentionally* pulling the wool over your eyes to get your money, but we urge you to be an investigator when it comes to your health. Here are some types of products to pay special attention to:

- Buttons and stickers
- Chips or diodes
- Pendants, amulets, and so on
- Harmonizers and neutralizers

Whatever products you consider, independent tests conducted by a certified independent laboratory are your best buying guide. Make sure the testing report shows effectiveness for both RF and ELF radiation blocking/reduction. Take extra time to research indirect and subjective tests, claims that may seem to lack basis in physics or scientific principles, or any other statements that you may consider questionable. Evaluate statements about how the products are designed. Does it make sense to you—or does it sound like some gobbledygook? If it does, it probably is!

Summary

Two out of the three most effective (and scientifically proven) methods for reducing your EMF exposure are completely free! You don't have to spend money on protection unless you choose to. We encourage you to do your own research and decide what's best for you and your family. There's no need to abandon the devices that have made our lives easier and, in many ways, more enjoyable—but it's also advisable to be smart about using them. Remember that there are simple steps you can take to reduce your exposure to a great degree.

Chapter 9

EMF SOURCES OF GREATEST CONCERN

"I do think that the amount of radiation exposure we get these days is exponentially higher than we did 15, 20 years ago. So anything you can do to limit your exposure to radiation is a good idea. But I don't think you need to give up any of these products. I try not to put the laptop directly on my lap, I try not to put the cell phone directly to my ear."

–Dr. Sanjay Gupta M.D., CNN chief medical correspondent

We now know that all electrical and electronic devices emit EMF radiation; it's just how they work. According to the laws of physics, to generate electricity, you must also create a magnetic field. So, because we live in a technologically advanced society, our exposure to EMFs is unavoidable.

With that said, it is also important to understand that some EMF sources emit more radiation than others, and that each can have a different effect on our bodies. A few common sources of EMF are listed here:

- **Power lines.** Highly elevated power lines carry very high voltages, yet may be considered relativity safe because typically humans know to maintain large distances between themselves and these lines.

- **Household wiring.** Lines running through our walls and ceilings emit low frequency EMFs and should be considered a potential health concern to electromagnetic hypersensitive people.
- **Electric blankets.** These warming devices can create an electromagnetic field that penetrate approximately 6–7 inches into the body; they have been linked to adverse events, including miscarriages and childhood leukemia, by experts such as Dr. Nancy Wertheimer PhD and physicist Ed Leeper.[157]
- **Microwave ovens.** Microwave radiation emissions are measured in milliwatts per centimeter squared (mW/cm^2) and are considered high-risk. The US safety limit on microwave emission exposure is 1 mW/cm^2. This limit is often exceeded when the body is close to older or faulty microwave ovens.
- **Alarms clocks.** These common timers produce very high EMFs, as much as 5–10 mG, up to three feet.
- **Hair dryers.** These styling tools can emit high levels of EMFs, but because of their intermittent use, exposure tends to be less harmful than other devices used over constant, long periods. Nonetheless, some EMF experts suggest that children should not use hair dryers because they may be sensitive to such exposure, as their brains and nervous systems are still in the developmental stage.

The "Big Five" Sources

We can't go into detail on every single source of EMF radiation, but in this chapter, we're going to explore what we call the "Big Five"—sources that are arguably the most immediate threats and making the largest impact on our daily lives. The Big Five are:

1. cell phones;
2. laptops;
3. tablet computers;
4. smart meters; and
5. Wi-Fi.

We'll give you some background on each source along with their potential health concerns. Then we'll give you some actions to take to limit exposure for you and your loved ones.

Cell Phones

"Given the extremely large number of people who use wireless communication devices, even a very small increase in the incidence of disease resulting from exposure to RFR resulting from those devices could have broad implications for public health."

–Dr. Michael Wyde et al., National Toxicology Program

Today, there is no electronic device more widespread than the cell phone. It has become the most rapidly adopted piece of consumer technology ever seen. The facts are staggering. Worldwide, there are upward of seven billion mobile phone users. According to the UN, more people have access to mobile phones than to working toilets![158] Recent Pew Research shows cell phone ownership among US adults has exceeded 90 percent.[159]

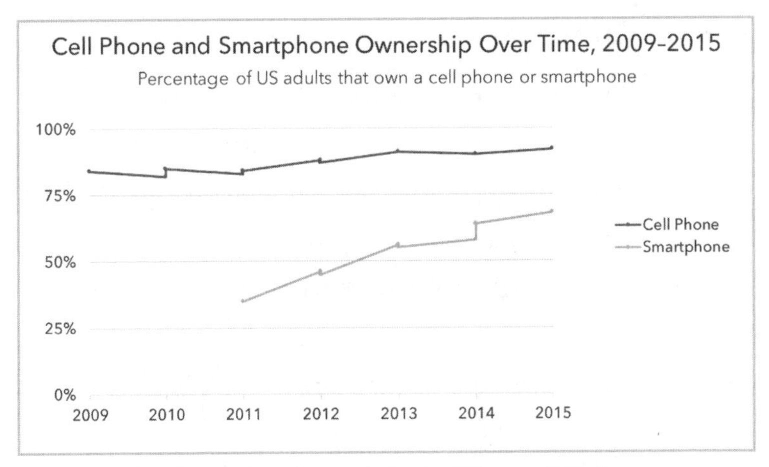

Data Source: Pew Research Center.[163] *Image: R. DeBaun.*

Since the advent of smartphones, which make everything from photography to video games possible at any time, it's rare we don't have them by our side. This has increased the unknown risks, because smartphones are basically mini computers. When our phones are turned on, their internal electronics produce ELF radiation. Separate from ELF, smartphones emit over-the-air RF radiation in the form of cellular, Wi-Fi, and Bluetooth transmissions. So, when you are using your phone, you are being exposed to several sources of ELF, and as many as three separate RF radiation sources.

Health Concerns

Studies are still ongoing to fully assess the potential long-term effects of mobile phone use, but as you've seen throughout the book, there is cause for concern. For example, a 2015 meta-analysis showed 93 percent of existing studies found that cell phone radiation has effects on biological systems.[161] And as mentioned previously in the book, the NTP study released in 2016 found that exposure to cell phone radiation can increase brain and heart tumors.

It is now generally accepted that cell phone radiation causes genetic changes and damage to DNA, resulting in the spread of abnormal cells. These effects cause illness, especially for those who use cell phones the most.

Actions

The following are a few suggestions for cell phone radiation safety:

- Use your cell phone only briefly against your head if possible.
- Use your speakerphone or headphones when appropriate.
- When you're not using your cell phone, keep it at least a foot away from your body.
- Consider using an EMF-shielding device on your phone that blocks both ELF and RF radiation.

Laptops

"Our data suggest that the use of a laptop computer wirelessly connected to the internet and positioned near the male reproductive organs may decrease human sperm quality."

–Dr. Conrado Avendaño, biochemist specializing in andrology and clinical research of sperm

Despite the name, did you know that laptops were not originally designed to be used in the lap? The first laptop computer, designed in 1979, actually weighed twenty-four pounds and was created to be placed on a desk.[162] As time passed, laptops became smaller, more portable, and started to be used primarily on the lap.

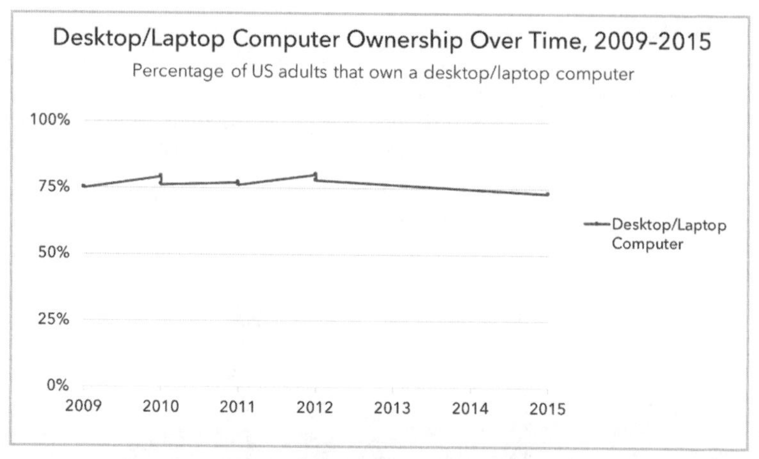

Data Source: Pew Research Center.[163] *Image: R. DeBaun.*

We all know that laptops, used directly in our laps, are very convenient—but they also expose us to the most intense sources of emissions extremely close to our reproductive organs, skin, and muscles.

When converting energy to perform their various functions, laptops generate ELF emissions. These fields radiate from the outer shell of laptops, created by internal sources (including processor activity, fans, hard drive operations, memory storage, and other computing functions).

To connect to the Internet and other devices, laptops transmit via Wi-Fi and Bluetooth transmissions, which are forms of RF radiation. When sending and receiving signals, laptops can radiate these emissions directly into your lap.

EMF expert and director of the Institute for Health and Environment at the University of Albany, Dr. David O. Carpenter, has said, "When you hold your laptop on your lap, what you're essentially doing is radiating your pelvis."[164]

When using laptops, we should think not only about EMF risks but about thermal ones as well. Many laptops (especially older models) generate a lot of heat from internal parts as well as have fans that exhaust hot air at temperatures exceeding 110°F. These fans keep the operating temperature stable inside the computer, ensuring proper performance, but the exhausted hot air exposes users to potentially harmful effects.

Health Concerns

Male infertility is a common concern, affecting about one in twenty adults. In many cases, a male can be infertile even though he is producing enough sperm. This happens due to defects in the sperm which prevents conception.[165] With this being the case, if you are male and looking to have children one day, you might want to rethink how often you use your laptop in your lap.

Research has shown that using a laptop directly in the lap, especially in conjunction with Wi-Fi, is associated with a decrease in sperm count and motility while causing sperm DNA damage.[166] Motility enables the sperm to swim to its intended target, the egg. Intact DNA allows the sperm to produce viable offspring. Obviously, sperm need both in order to successfully impregnate a healthy egg.

Because of the growing body of research, all pointing in the same direction, it is considered advisable to avoid exposure to EMF radiation emitted from the bottom of your laptop. Minimizing direct exposure is essential to preventing unnecessary reproductive complications.

Actions

The following are a few suggestions for laptop radiation safety:

- When using your laptop, place it on a table or desk, if possible, instead of directly in your lap.
- Turn off your Wi-Fi and Bluetooth when you are not using them.
- It is best not to use a laptop on a pillow or other object as a buffer for your lap. Although this may create some distance (which helps reduce exposure), it will not eliminate it. A pillow or other object may also disrupt the flow of air, which is essential to cool the laptop and keep it from overheating.
- If you plan to place your laptop in your lap, use a shield that reduces or eliminates both ELF and RF radiation as well as heat.

Tablet Computers

"I think we should use the airplane mode on our phones and tablets. Just putting it on that reduces a lot of radiation."

–Dr. Harold L. Naiman, MD, pediatrician

The tablet computer is an incredible device and one of the most useful tools of our era. We can use it to do everything from watch a movie or play a game to search the Internet or talk face-to-face with a loved one halfway across the world. It's amazing to think how we lived without them.

After several types of tablets were universally snubbed by the masses, many credit the explosion of tablets' popularly with the release of the iPad in 2010.[167] Tablets are so new that they have taken the consumer market by storm. Year after year, their popularity just keeps growing, to the extent that they are even replacing the laptop. In 2015, tablet sales actually surpassed that of desktop PCs and laptops combined.[168]

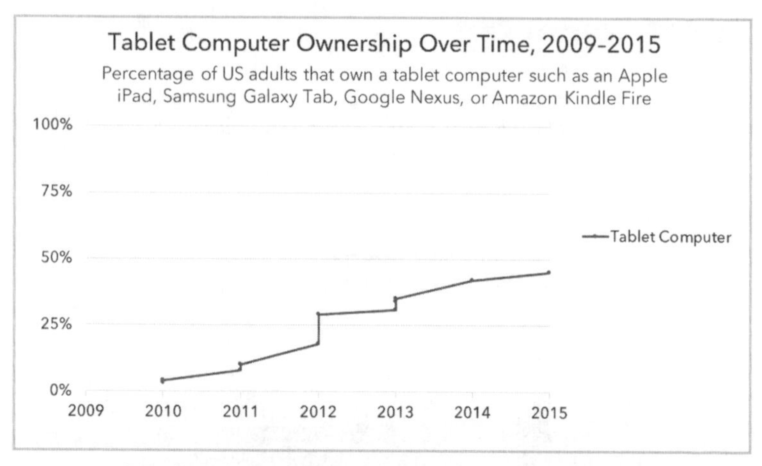

Data Source: Pew Research Center.[169] *Image: R. DeBaun.*

As with any popular electronic device, we run the risk of equating popularity with safety. We give tablets to our young and have started incorporating them into our schools. We use tablets in our homes and at work. Tablets are so small, convenient, and easy to use, yet they are powerful enough to replace other devices we use. We end up using them all the time because they are so versatile. Yet that is where the concern comes in.

Even though tablets are very compact, they are still packed with powerful hardware that includes powerful microprocessors, backplanes, batteries, and other electronic components. When turned on, they consume power and give off, as a by-product, ELF radiation.

Separately, tablets transmit and receive signals over cellular (if your tablet is equipped with this feature), Bluetooth, and/or Wi-Fi RF connections. So, unless you turn off your tablet's Wi-Fi or place it in airplane mode, your device will be constantly transmitting and receiving RF signals.

While tablet computers are coming out in smaller and thinner lightweight models that look pretty harmless, they may lead to some unintended side effects.

Health Concerns

Since they have only become popular in the past few years, research focused specifically on the health risks associated with tablets is extremely limited. Most research has been focused on more-established devices, such as laptops and cell phones. With the exponential growth of tablets, though, we expect to see more specific research in time.

However, given the available research related to EMFs already discussed in this book, we can assume that tablets pose similar risks as laptops, cell phones, and other electronic devices. Tablet computers have several sources of radiation, including both wireless and cellular signals, as well as ELF radiation. People use tablets either very close to their bodies or directly on their bodies for long periods of time (probably much longer than using a cell phone against the head). And it is even worse for children, who are more susceptible. Over two-thirds of children under twelve years old living in a tablet-owning household have used the tablet device.[170] And this number will only continue to grow.

So in our opinion, and according to related research, this risk of tablet radiation is very real, and precautionary steps should be made to limit exposure.

Actions

The following are a few suggestions for tablet radiation safety:

- Avoid using your tablet directly on your lap or holding it close to your body. Work with it as far away from your body as possible, preferably resting on a desk, tabletop, or other surface. Note that placing your tablet on a pillow or other object will not reduce radiation, but it will create distance from the body, which helps to reduce exposure.
- When using a tablet, always turn it to airplane mode when you are not using the Internet. If the Wi-Fi setting is left active, it will continue to emit radiation even when you're not connected to a Wi-Fi network.
- Think about using an EMF shield that blocks both ELF and RF radiation.

Smart Meters

"Over the past two years, there has been mounting medical and scientific evidence of the grave biological dangers to humans from so-called "Smart" Meters exposure that are being installed by the hundreds of thousands all over North America and Europe. Scientists have been documenting the EMF/RF exposure effects for decades. However, it is only in the last two years, with the constant wireless Electromagnetic Radiation exposure to these new meters, that other medical evidence (down to the cellular level) has been reported."

–Dr. Ilya Sandra Perlingieri, environmental activist

The days of someone coming around to each house on your block and manually reading all the gas, electric, and water meters is quickly disappearing. Instead, utility companies are spending billions of dollars to upgrade their technology to replace traditional meters with so-called smart meters that automatically send usage information back to the company. In fact, during a four-year period, from 2010 through 2013, the electric industry spent an estimated total of $18 billion deploying smart-grid technology in the United States.[171] As of 2014, about 58.5 million smart meters had already been installed in the United States, 88 percent of which were for residential customer installations.[172]

In general, these smart meters send data to the utility company via RF signals from fifty to thousands of times a day, depending on who you ask. Each pulse can last from less than a second to twelve or more hours, depending on the action being performed. The smart meter RF transmitter frequency of operation is typically in the 900 MHz and 2.4 GHz bands. Power output is normally around 1 watt in the 902 MHz band and much less in the 2.4 GHz band.

Smart meters allow even more radiation into our already-saturated daily lives. Think of how many of these meters are transmitting within one block of your house. Don't forget, these are additional EMF emissions that penetrate your home, where you likely already have routers, computers, and cell phones. If you live in an apartment building, their affect is even greater.

The smart meter issue is troubling for many reasons besides just EMF emissions. First, companies don't always get consent from customers to replace their old meters with these new smart meters. Second, there are also issues of privacy, as they can act as a means of surveillance when they collect data about usage. For instance, California utility companies have admitted to sharing this data with the government as well as third parties.[173]

Then there is the issue of security. Hackers can easily access this new grid. Former CIA director James Woolsey even called it "a really, really stupid grid."[174] Hacking vulnerabilities leave the power grid open to viruses and other threats that could shut it down or destroy it completely.

Health concerns

Unfortunately, there has been little to no study on the health effects of smart meters. However, a handful of surveys have been conducted as of this time of the publication of this book. For example, when Victoria, Australia, commissioned a rollout of smart meters in 2006, suddenly an entire population couldn't avoid human-made RF radiation. Over 142 people reported adverse health effects from the wireless smart meters. The most frequently reported symptoms were insomnia, headaches, tinnitus (ringing in the ears), fatigue, cognitive disturbances, dysesthesias (abnormal sensations), and dizziness.[175]

In 2011, an online survey was commissioned by the EMF Safety Network to investigate public health and safety complaints about wireless utility smart meters. In the survey, respondents reported a variety of symptoms from sleep problems to high blood pressure.[176]

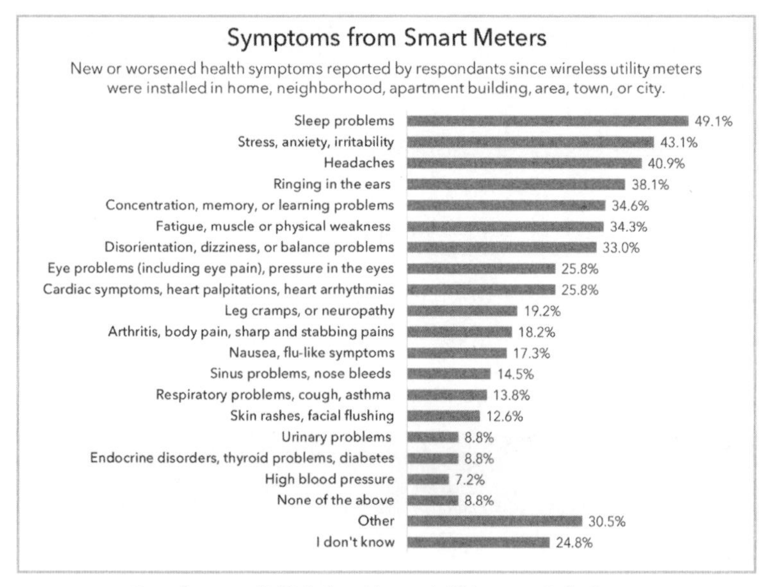

Data Source: EMF Safety Network.[177] Image: R. DeBaun.

Once again, we don't know definitively the long-term effects that smart meters will have, but we do have a clue. Many studies have already shown that RF radiation can have serious side effects, and, once again, the World Health Organization says that smart meter RF signals may cause cancer, placing them in the same category as lead and DDT. What more will we find in ten, twenty, or thirty years if this issue is left unchecked?

Actions

The following are a few suggestions for smart meter safety:

- Have the smart meter replaced with a meter that does not emit RF radiation, if possible.
- Place shielding on the inside wall of your home where the smart meter is located to block the RF signal from entering.
- Although less effective, you may consider using EMF-blocking paint on the interior wall opposite the smart meter.

Wi-Fi Networks

"As a scientist, I'm a little bit surprised that politicians, they take these types of chances with the whole population, and the population has never really been informed. No one has asked you if you want to be whole body radiated 24 hours around the clock, every day of the week."

- Dr. Olle Johansson, PhD, associate professor, Experimental Dermatology Unit, Department of Neuroscience, Karolinska Institute, Stockholm, Sweden

Our world is becoming increasingly connected via Wi-Fi. By the end of 2015, 752 million households worldwide were connected to the Internet via fixed-line broadband. Of those households, 521 million (or 69 percent) used a wireless router to create a Wi-Fi home network.[178] But it's not just homes where Wi-Fi is growing. More cities, schools, hotels, stores, stadiums, and just about everywhere else are becoming Wi-Fi–enabled. Even a smart phone can be its own Wi-Fi hot spot. But with the rush to build these networks, it's crucial to consider the potential downsides.

A Wi-Fi network's principal device is a router that transmits RF signals at 2.4 GHz or 5.0 GHz. Routers are switching boxes, programmed to send data where it needs to go, while enabling Internet connectivity. Wi-Fi is typically used in a home, office, business, or public space and is usually called a wireless local area network. You can recognize a router because there is a bank of about four to eight jacks for plugging in Ethernet cables, while the front sports an array of blinking LED lights, indicating the status of data traffic.

The 5.0 GHz channel transmissions are potentially more harmful than the 2.4 GHz channel transmissions, given that the frequency of waves is double per unit of time. Referred to as "carrier waves," each of these waves have pulsing frequencies within them, transporting "information content." The 5.0

GHz channel supports much faster download speeds because there are more packets of information being transmitted.

The amount of Wi-Fi networks today and the fact they are always on raises concerns. Even if you live in a semi-populated area, you still get long lists of Wi-Fi networks you can connect to. We are basically saturated by RF signals 24-7.

Health Concerns

The rapid spread of electronic communication around the world has raised public health concerns about the dangers of long-term environmental EMF radiation exposure. They have shown Wi-Fi can affect tissues, fertility, brain electrical activity, nervous systems, cognitive function, sleep, heart rate, and blood pressure—yet all within the FCC's exposure guidelines.

As Dr. Martin Blank of Columbia University wrote to a school board concerning a newly installed Wi-Fi network, "Compliance with FCC guidelines, unfortunately, is not in any way an assurance of safety today, as the guidelines are fundamentally flawed." He went on to say, "Until the guidelines and advisories in the United States are updated, the intelligent thing for your Board of Trustees to do is to exercise the Precautionary Principle and hard wire all internet connections."[179]

Echoing his advice, is best to use caution when it comes to Wi-Fi and limit exposure if possible.

Actions

The following are a few suggestions for Wi-Fi radiation safety:

- Use Ethernet cables to create a hardwired network instead of using over-the-air RF transmissions.
- Turn off your router when not in use.
- Place the router at least ten feet or more away from where people are active.
- Do not put wireless routers near bedrooms.
- For children, avoid using Wi-Fi in schools, nurseries, day care centers, playrooms, and so on.
- It is always best to place a router in a location that is the least occupied part of the home or building to minimize constant exposure.

Chapter 10

FINAL THOUGHTS: STAYING SAFE IN THE MODERN WORLD

"The biggest problem we have is that we know most environmental factors take several decades of exposure before we really see the consequences."

–Dr. Keith Black, chairman of neurology at Cedars-Sinai Medical Center, Los Angeles

"Really at the heart of my concern is that we shouldn't wait for a definitive study to come out, but err on the side of being safe rather than sorry later."

–Dr. Ronald B. Herberman, University of Pittsburg Cancer Institute

Quality of life can be improved by technology, but each family should also weigh the risks. Photo: © istockphotos.com/squaredpixels.

If you were offered a tool to make your life easier, but in exchange you had to give up some of your lifespan, would you take it? What if you had to give up your child's lifespan, health, or well-being? What about your chance to have children? Even if your risk were relatively small—say, a 3 percent chance of developing a glioblastoma—would you take it? (NTP 2016). This is not a hypothetical question. This is what is going on right now.

Our lives have become so integrated with technology that we don't stop to think about what we are doing. We just introduce more and more electronic devices into our lives, but we don't seem to consider the potential implications that stem from our dependence on them. These devices all emit EMF radiation, which has measurably demonstrated an effect on living organisms by damaging DNA, cells, and living tissues. Surrounding ourselves with these devices, yet choosing to remain ignorant of how they work, is like playing Russian roulette, blindfolded, with multiple guns (any of which may be loaded).

Electronic devices bring an amazing amount of convenience to our lives because of their mobile communication; however, research shows they can also bring risk. EMF exposure can be harmful if people are unaware of its hazards. These hazards are unnecessary gambles because they are avoidable ones.

Devices can be turned off when not needed and held away from our bodies when in use. We can limit the time we keep our cell phones in our pockets. We can disable our Wi-Fi. We can prevent our young from using gadgets before they reach a certain age. We can even shield our bodies from too much exposure. There are numerous ways we can protect ourselves.

We only put ourselves in jeopardy by opting to remain passive or ignorant about this information. Our response, as a discerning public, should not be to panic or act as victims; we are not victims. We should just pay heed to what science is saying and take the necessary actions to protect ourselves and our loved ones.

EMF radiation has the power to transform our lives—for better or worse—so we need to stay abreast of current knowledge and know what we surround ourselves with as well as the impact it has. Our well-being, livelihood, and lifespan are all affected. We can still use and enjoy our technology if we take the right measures to use it correctly—at appropriate distances and frequencies and with precaution.

NEXT STEPS

Obviously, things move fast these days and information is constantly changing. If you want to stay up-to-date, we recommend that you visit the *Radiation Nation* website regularly.

www.RadiationNationBook.com

We also urge you to connect with our community at the following web address:

www.RadiationNationBook.com/Connect

Here, you can sign up for the *Radiation Nation* newsletter to get insights into the latest EMF research and scientific studies. We will also share more key principles for simple and effective EMF safety as well as practical tips to protect yourself and your loved ones.

OUR THANKS TO YOU

We hope you enjoyed this book and, at the very least, are more aware of what you can do to protect yourself and your loved ones from EMF exposure. Helping others is our greatest mission, and we are honored to be able to share this information with you.

If this book helped or at least informed you in any way, we would be extremely grateful if you could take a moment to write an honest, sincere review of this book. It will only take a few minutes, and it would help us out more than you can imagine!

Sincerely,

Daniel and Ryan

ABOUT THE AUTHORS

Daniel T. DeBaun

Daniel is an internationally recognized and influential expert in in EMF radiation, related health issues, and shielding, with a particular focus on the effect of exposure from personal mobile devices. Daniel's concern regarding the health impact of EMF emissions grew from over thirty years of engineering experience in the telecommunication industry, where he held a variety of leadership and executive positions at SAIC, Telcordia, AT&T, and Bell Labs.

Through the course of his career, Daniel has created requirements for large telecommunication systems, led technical divisions responsible for establishing industry standards, and formed analysis adherence testing for next-generation digital transmission systems. Daniel also oversaw laboratories that analyzed electromagnetic radiation interference, electrical signals, and digital formats.

Daniel is the inventor of DefenderShield®, a form of protection technology for modern electronic devices. In addition to his work with DefenderShield®, Daniel is a highly regarded industry consultant, writer, and speaker, as well as a frequent guest on national radio and television programs discussing the impacts of EMF radiation and protection options.

As co-author of Radiation Nation and co-founder of DefenderShield®, Ryan is considered a leading expert in EMF radiation safety issues.

Throughout his extensive career, Ryan has worked across all aspects of media with companies including HBO, CNBC, CBS, ABC and has been instrumental in taking complex subject matter and making it it accessible for the general public.

His media and communications background, as well as his in-depth understanding of health and wellness, allows him to help educate worldwide communities about the potential concerns of EMF radiation exposure and ways to minimize the risks.

In his free time, Ryan enjoys yoga, art, and educational activities.

Appendix A

EMF PROTECTION TIP SHEET

- Use your cell phone only briefly against your head. You can also use speakerphone or headphones.
- When you're not using your cell phone, keep it at least a foot away from your body.
- Limit the use of laptops and tablets directly on your lap. For longer use, consider using an EMF shield.
- Use an Ethernet cable to connect to the Internet instead of Wi-Fi, if possible. (Make sure to go into your computer settings and turn off the Wi-Fi and Bluetooth when you don't need them.)
- If you must use Wi-Fi, turn on your router only when needed and turn it off when you're finished.
- Make sure that your wireless router is at least ten feet (or more) away from where you are sitting.
- For both desktop computers and laptops, use a wired mouse and keyboard if possible.
- Push your desktop monitor away from you, and place your laptop on a table. Although there is relatively low radiation, just increasing the distance between you and the electronic device cuts down on exposure.
- If you are a parent, please be advised that children are more susceptible to EMF emissions than adults. Limit your child's use of mobile devices, particularly when used close to their bodies.
- Remember, time and distance are your friends! Always limit the amount of time you spend using your electronic devices. The more distance you have between your body and the device, the better.

Appendix B
HELPFUL RESOURCES

The following websites offer an array of EMF safety–related information and resources:

www.bioinitiative.org

www.cancer.gov

www.citizensforsafetechnology.co

www.electromagnetichealth.org

www.emwatch.com

www.emfpolicy.org

www.saferemf.com

www.icnirp.org

www.kawarthasafetechnology.org

www.microwavenews.com

www.powerwatch.org.uk

www.safeinschool.org

www.safeschool.ca

www.wiredchild.org

www.who.int/peh-emf

Suggested websites should not be taken as an endorsement or a recommendation of any third-party products or services offered by virtue of any information, material, or content linked from or to this site. The opinions expressed in sites do not necessarily represent our position and are provided for information purposes only.

WORKS CITED

Introduction

1 Marguerite Reardon, "The Trouble with the Cell Phone Radiation Standard," CNET, June 2, 2011, http://www.cnet.com/news/the-trouble-with-the-cell-phone-radiation-standard/.

2 Kenneth Chang, "Debate Continues on Hazards of Electromagnetic Waves," *New York Times*, July 7, 2014, http://www.nytimes.com/2014/07/08/science/debate-continues-on-hazards-of-electromagnetic-waves.html.

3 Geoffrey Kabat, "The *New York Times* Revisits the 'Debate' Over Electromagnetic Fields, Reviving Baseless Fears, While Ignoring What Has Been Learned," *Forbes*, July 8, 2014, http://www.forbes.com/sites/geoffreykabat/2014/07/08/the-new-york-times-revisits-the-debate-over-electromagnetic-fields-reviving-baseless-fears-while-ignoring-what-has-been-learned/#1294a9a13021.

4 Nick Bilton, "The Health Concerns in Wearable Tech," *New York Times*, March 18, 2015, http://www.nytimes.com/2015/03/19/style/could-wearable-computers-be-as-harmful-as-cigarettes.html.

5 Jeromy Johnson, "Wireless Empire Strikes Back," Protect Your Family from EMF Pollution, March 21, 2015, https://www.emfanalysis.com/wireless-empire-strikes-back/.

6 Nick Bilton, "The Health Concerns in Wearable Tech," *New York Times*, March 18, 2015, http://www.nytimes.com/2015/03/19/style/could-wearable-computers-be-as-harmful-as-cigarettes.html.

Chapter 1

7 Tom Geoghegan, "Twitter, Telegram and E-mail: Famous First Lines," *BBC Magazine*, March 21, 2011, http://www.bbc.com/news/magazine-12784072.

8 Graham Smith, "The Day the Mobile Phone Went Public 38 Years ago, Leaving New Yorkers Bemused and Bewildered," DailyMail.com, last modified April 5, 2011, http://www.dailymail.co.uk/sciencetech/article-1373272/The-day-Martin-Cooper-took-mobile-phone-public-leaving-New-Yorkers-bemused-bewildered.html.

9 Larry Seltzer, "Cell Phone Inventor Talks of First Cell Call," InformationWeek, April 3, 2013, http://www.informationweek.com/wireless/cell-phone-inventor-talks-of-first-cell-call/d/d-id/1109376?.

10 "Marty Cooper Interview for Scene World Magazine," YouTube video, 1:29:01, posted by "TMCrole," February 12, 2015, https://www.youtube.com/watch?v=B6OKTJMavtw&feature=youtu.be&list=PLE3A053CEEE38AE81.

11 "Martin Cooper and the History of Cell Phone," About.com Money, last modified October 31, 2016, http://inventors.about.com/cs/inventorsalphabet/a/martin_cooper.htm.

12 "Key ICT Indicators for Developed and Developing Countries and the World (Totals and Penetration Rates)," International Telecommunication Union (ITU), 2015, http://www.itu.int/en/ITU-D/Statistics/Documents/statistics/2015/ITU_Key_2005-2015_ICT_data.xls.

13 "Number of Mobile (Cellular) Subscriptions Worldwide From 1993 to 2016 (in Millions)," Statista, 2016, http://www.statista.com/statistics/262950/global-mobile-subscriptions-since-1993/.

14 "Electromagnetic Fields and Public Health: Mobile Phones," World Health Organization, Oct. 2014, http://www.who.int/mediacentre/factsheets/fs193/en/.

15 "Device Ownership Over Time," Pew Research Center, http://www.pewinternet.org/data-trend/mobile/device-ownership/.

16 Nielsen, *The Total Audience Report*, Q4 2015, http://www.nielsen.com/content/dam/corporate/us/en/reports-downloads/2016-reports/q4-2015-total-audience-report.pdf.

17 Ibid.

18 Carol Pogash, "Cell Phone Ordinance Puts Berkley at Forefront of Radiation Debate," *New York Times*, July 21, 2015, http://www.nytimes.com/2015/07/22/us/cellphone-ordinance-puts-berkeley-at-forefront-of-radiation-debate.html.

19 "France: New National Law Bans WIFI in Nursery School," Environmental Health Trust, January 19, 2015, http://ehtrust.org/france-new-national-law-bans-wifi-nursery-school/.

20 Next-Up Organization, *France National Library Gives-Up Wi-Fi*, July 4, 2008, http://www.next-up.org/pdf/FranceNationalLibraryGivesUpWiFi07042008.pdf.

21 "EHT Lauds Israel's Ban on Wi-Fi in Kindergarten in Schools Additionally, Italian State of Tyrol Calls for Limiting Wireless in Schools," Environmental Health Trust, http://ehtrust.org/eht-lauds-israels-ban-on-wi-fi-in-kindergarten-and-limits-to-childrens-wireless-exposures-in-schools-additionally-the-italian-state-of-tyrol-now-calls-for-limiting-wireless-in-schools/.

22 Shreya Shah, "Radiation Panic Grips Mumbai," *India Real Time* (blog), *Wall Street Journal*, February 25, 2013, http://blogs.wsj.com/indiarealtime/2013/02/25/radiation-panic-grips-mumbai/.

23 "EHT Lauds Israel's Ban on Wi-Fi in Kindergarten in Schools Additionally, Italian State of Tyrol Calls for Limiting Wireless in Schools," Environmental Health Trust, http://ehtrust.org/eht-lauds-israels-ban-on-wi-fi-in-kindergarten-and-limits-to-childrens-wireless-exposures-in-schools-additionally-the-italian-state-of-tyrol-now-calls-for-limiting-wireless-in-schools/.

24 David Gee, "Late Lessons from Early Warnings: Towards Realism and Precaution with EMF?", *Pathophysiology* 16, nos. 2–3 (2009): 217–231, doi:http://dx.doi.org/10.1016/j.pathophys.2009.01.004.

25 Ashley Csanady, "Local Teachers Unions in Ontario Call for Moratorium on Wi-Fi Use in Schools," *National Post*, March 21, 2016, http://news.nationalpost.com/news/canada/local-teachers-unions-in-ontario-latest-to-call-for-moratorium-on-wi-fi-use-in-schools.

26 Eleanor Abaya and Fred Gilbert, "Lakehead Says No to Wi-Fi," *Lakehead University Magazine*, Spring/Summer 2008, http://magazine.lakeheadu.ca/page.php?p=81.

27 Joel Moskowitz, "Drs. Oz and Gupta Caution About Cell Phones," *Electromagnetic Radiation Safety* (blog), January 16, 2013, http://www.saferemr.com/2013/01/comments-drs-oz-and-gupta-call-for.html.

28 Carlo Dellaverson, "Witnessing Papal History Changes with Digital Age," NBC News, March 14, 2013, http://photoblog.nbcnews.com/_news/2013/03/14/17312316-witnessing-papal-history-changes-with-digital-age.

29 Ibid.

Chapter 2

30 US Army Intelligence and Security Command, *Bioeffects of Selected Non-Lethal Weapons*, February 17, 1998, https://www.wired.com/images_blogs/dangerroom/files/Bioeffects_of_Selected_Non-Lethal_Weapons.pdf.

31 "Cell Phones and Cancer Risk," National Cancer Institute, last modified May 27, 2016, http://www.cancer.gov/cancertopics/causes-prevention/risk/radiation/cell-phones-fact-sheet.

32 "Electromagnetic Fields and Public Health: Mobile Phones," World Health Organization, Oct. 2014, http://www.who.int/mediacentre/factsheets/fs193/en/.

33 Martin L. Pall, "Microwave Frequency Electromagnetic Fields (EMFs) Produce Widespread Neuropsychiatric Effects Including Depression," *Journal of Chemical Neuroanatomy* 75, pt. B (2016): 43–51, doi:http://dx.doi.org/10.1016/j.jchemneu.2015.08.001.

34 Nora D. Volkow et al., "Effects of Cell Phone Radiofrequency Signal Exposure on Brain Glucose Metabolism," *JAMA* 305, no. 8 (2011): 808–813, doi:10.1001/jama.2011.186.

Chapter 3

35 "Bioinitiative 2012 Media," BioInitiative Working Group, http://www.bioinitiative.org/media/.

36 Marguerite Reardon, "The Trouble with the Cell Phone Radiation Standard," CNET, June 2, 2011, http://www.cnet.com/news/the-trouble-with-the-cell-phone-radiation-standard/.

37 Daisy Yuhas, "It's Electric: Biologists Seek to Crack Cell's Bioelectric Code," *Scientific American*, March 27, 2013, http://www.scientificamerican.com/article/bioelectric-code/.

38 Martin Blank and Reba Goodman, "Electromagnetic Fields Stress Living Cells," *Pathophysiology* 16, nos. 2–3 (2009): 71–78, doi:http://dx.doi.org/10.1016/j.pathophys.2009.01.006.

39 Z. Somosy, "Radiation Response of Cell Organelles," *Micron* 31, no. 2 (2000): 165–181, doi:http://dx.doi.org/10.1016/S0968-4328(99)00083-9.

40 Michele Caraglia et al., "Electromagnetic Fields at Mobile Phone Frequency Induce Apoptosis and Inactivation of the Multi-Chaperone Complex in Human Epidermoid Cancer Cells," *Journal of Cellular Physiology* 204, no. 2 (2005): 539–548, doi:10.1002/jcp.20327.

41 "Mobile Phone 'Brain Risk'," BBC News, November, 6, 1999, http://news.bbc.co.uk/2/hi/health/507112.stm.

42 Suzanne Clancy, "DNA Damage and Repair: Mechanisms for Maintaining DNA Integrity," *Nature Education* 1, no. 1 (2008): 103, http://www.nature.com/scitable/topicpage/dna-damage-repair-mechanisms-for-maintaining-dna-344.

43 "EMFs in the Workplace," National Institute for Occupational Safety and Health, last modified June 6, 2014, http://www.cdc.gov/niosh/docs/96-129/.

44 Nora D. Volkow et al., "Effects of Cell Phone Radiofrequency Signal Exposure on Brain Glucose Metabolism," JAMA 305, no. 8 (2011): 808–813, doi:10.1001/jama.2011.186.

45 Henry Lai and Lennart Hardell, "Cell Phone Radiofrequency Radiation Exposure and Brain Glucose Metabolism," *JAMA* 305, no. 8 (2011): 828–829, doi:10.1001/jama.2011.201.

46 "Glioblastoma," Wikipedia, last modified October 25, 2016, https://en.wikipedia.org/wiki/Glioblastoma_multiforme.

47 Melinda Wenner, "Fact or Fiction?: Cell Phones Can Cause Brain Cancer," *Scientific American*, November 21, 2008, http://www.scientificamerican.com/article/fact-or-fiction-cell-phones-can-cause-brain-cancer/.

48 G. Zada et al., "Incidence Trends in the Anatomic Location of Primary Malignant Brain Tumors in the United States: 1992–2006," *World Neurosurgery* 77, nos. 3–4 (2012): 518–524, doi:10.1016/j.wneu.2011.05.051.

49 Devra Davis, Anthony B. Miller, and L. Lloyd Morgan, "Why There Can Be No Increase in All Brain Cancers Tied with Cell Phone Use," *OUPblog* (blog), May 16, 2016, http://blog.oup.com/2016/05/brain-cancers-cell-phone-use/.

50 Lennart Hardell and Michael Carlberg, "Mobile Phone and Cordless Phone Use and the Risk of Glioma—Analysis of Pooled Case-Control Studies in Sweden, 1997–2003 and 2007–2009," *Pathophysiology* 22, no. 1 (2015): 1–13, doi:http://dx.doi.org/10.1016/j.pathophys.2014.10.001.

51 Dina Fine Maron, "Major Cell Phone Radiation Study Reignites Cancer Questions," *Scientific American*, May 27, 2016, http://www.scientificamerican.com/article/major-cell-phone-radiation-study-reignites-cancer-questions/.

52 "Cell Phones," National Toxicology Program, last modified September 13, 2016, http://ntp.niehs.nih.gov/results/areas/cellphones/.

53 "Cell Phone Radiation Boosts Cancer Rates in Animals; $25 Million NTP Study Finds Brain Tumors," Microwave News, May 25, 2016, http://microwavenews.com/news-center/ntp-cancer-results.

54 Joel Moskowitz, "National Toxicology Program Finds Cell Phone Radiation Causes Cancer," *Electromagnetic Radiation Safety* (blog), September 7, 2016, http://www.saferemr.com/2016/05/national-toxicology-progam-finds-cell.html.

55 Michael Wyde et al., "Report of Partial Findings from the National Toxicology Program Carcinogenesis Studies of Cell Phone Radiofrequency Radiation in Hsd: Sprague Dawley® SD Rats (Whole Body Exposure)," (working paper, National Institute of Environmental Health Sciences, Research Triangle Park, North Carolina, 2016), doi:http://dx.doi.org/10.1101/055699.

56 "Electromagnetic Fields and Public Health: Mobile Phones," World Health Organization, Oct. 2014, http://www.who.int/mediacentre/factsheets/fs193/en/.

57 "ACS Responds to New Study Linking Cell Phone Radiation to Cancer," American Cancer Society, May 2017, http://pressroom.cancer.org/NTP2016.

Chapter 4

58 "The Reports of the Surgeon General," U.S. National Library of Medicine, https://profiles.nlm.nih.gov/ps/retrieve/Narrative/NN/p-nid/60.

59 De-Kun Li, Hong Chen, and Roxana Odouli, "Maternal Exposure to Magnetic Fields During Pregnancy in Relation to the Risk of Asthma in Offspring," *JAMA* 165, no. 10 (2011): 945–950, doi:10.1001/archpediatrics.2011.135.

60 Martha R. Herbert and Cindy Sage, "Autism and EMF? Plausibility of a Pathophysiological Link Part II," *Pathophysiology*

20, no. 3 (2013): 211–234, doi:http://dx.doi.org/10.1016/j.pathophys.2013.08.002.

61 R. C. Kane, "A Possible Association Between Fetal/Neonatal Exposure to Radiofrequency Electromagnetic Radiation and the Increased Incidence of Autism Spectrum Disorders (ASD)," *Medical Hypotheses* 62, no. 2 (2004): 195–197, doi:10.1016/S0306-9877(03)00309-8.

62 M. Blank, "Cell Biology and EMF Safety Standards," *Electromagnetic Biology and Medicine* 34, no. 4 (2015): 387–389, doi:10.3109/15368378.2014.952433; M. Blank and R. M. Goodman, "Electromagnetic Fields and Health: DNA-Based Dosimetry," *Electromagnetic Biology and Medicine* 31, no. 4 (2012): 243–249, doi:10.3109/15368378.2011.624662; S. Dasdaq et al., "Effects of 2.4 GHz Radio Frequency Radiation Emitted from Wi-Fi Equipment on MicroRNA Expression in Brain Tissue," *International Journal of Radiation Biology* 97, no. 7 (2015): 555–561, doi:10.3109/09553002.2015.1028599; P. W. French, M. Donnellan, and D. R. McKenzie, "Electromagnetic Radiation at 835 Mhz Changes the Morphology and Inhibits Proliferation of a Human Astrocytoma Cell Line," *Bioelectrochemistry and Bioenergetics* 43, no. 1 (1997): 13–18, doi:10.1016/S0302-4598(97)00035-4; S. Gulati et al., "Effect of GSTM1 and GSTT1 Polymorphisms on Genetic Damage in Humans Populations Exposed to Radiation from Mobile Towers," *Archives of Environmental Contamination and Toxicology* 70, no. 3 (2016): 615–625, doi:10.1007/s00244-015-0195-y; Y. Z. Li et al., "Effects of Electromagnetic Radiation on Health and Immune Function of Operators," *Chinese Journal of Industrial Hygiene and Occupational Diseases* 31, no. 8 (2013): 602–605, http://www.ncbi.nlm.nih.gov/pubmed/24053963; C. T. Mihai et al., "Extremely Low-Frequency Electromagnetic Fields Case DNA Strand Breaks in Normal Cells,"

Journal of Environmental Health Science and Engineering 12, no. 1 (2014): 15, doi:10.1186/2052-336X-12-15.

63 G. Mihailović, et al., "Epidemiological Features of Brain Tumors," *Srpski arhiv za celokupno lekarstvo* 141, nos. 11–12 (2013): 823–829, http://www.ncbi.nlm.nih.gov/pubmed/24502107/.

64 M. C. Turner et al., "Occupational Exposure to Extremely Low-Frequency Magnetic Fields and Brain Tumor Risks in the INTEROCC Study," *Cancer Epidemiology, Biomarkers and Prevention* 23, no. 9 (2014): 1863–1872, doi:10.1158/1055-9965.EPI-14-0102.

65 Lennart Hardell and Michael Carlberg, "Mobile Phone and Cordless Phone Use and the Risk of Glioma—Analysis of Pooled Case-Control Studies in Sweden, 1997–2003 and 2007–2009," Pathophysiology 22, no. 1 (2015): 1–13, doi:http://dx.doi.org/10.1016/j.pathophys.2014.10.001.

66 Alexander Lerchl et al., "Tumor Promotion by Exposure to Radiofrequency Electromagnetic Fields Below Exposure Limits for Humans," *Biochemical and Biophysical Research Communications* 459, no. 4 (2015): 585–590, doi:http://dx.doi.org/10.1016/j.bbrc.2015.02.151.

67 John G. West et al., "Multifocal Breast Cancer in Young Women with Prolonged Contact between Their Breasts and Their Cellular Phones," *Case Reports in Medicine* (2013), doi:10.1155/2013/354682.

68 G. Zhao et al., "Relationship Between Exposure to Extremely Low-Frequency Electromagnetic Fields and Breast Cancer Risk: A Meta-Analysis," *European Journal of Gynaecological Oncology* 35, no. 3 (2014): 264–269, http://www.ncbi.nlm.nih.gov/pubmed/24984538.

69 Ibid.

70 I. Calvente et al., "Exposure to Electromagnetic Fields (Non-Ionizing Radiation) and Its Relationship with Childhood Leukemia: A

Systematic Review," *Science of The Total Environment* 408, no. 16 (2010): 3062–3069, doi:http://dx.doi.org/10.1016/j.scitotenv.2010.03.039.

71 "Childhood Leukaemia," EMFs.info, http://www.emfs.info/health/leukaemia/.

72 Michael Kundi, *Evidence for Childhood Cancers (Leukemia)* (2012), http://www.bioinitiative.org/report/wp-content/uploads/pdfs/sec12_2012_Evidence_%20Childhood_Cancers.pdf.

73 "Electromagnetic Fields and Public Health," World Health Organization, Dec. 2005, http://www.who.int/peh-emf/publications/facts/fs296/en/.

74 Badereddin Mohamad Al-Ali et al., "Cell Phone Usage and Erectile Function," *Central European Journal of Urology* 66, no. 1 (2013): 75–77, doi:10.5173/ceju.2013.01.art23.

75 A. Petrova, "Risk of Radiation Exposure to Children and Their Mothers," in *Encyclopedia of Environmental Health* (2011): 878–886, doi:http://dx.doi.org/10.1016/B978-0-444-52272-6.00212-9.

76 De-Kun Li et al., "A Prospective Study of *In-Utero* Exposure to Magnetic Fields and the Risk of Childhood Obesity," *Scientific Reports* 2, article no. 540 (2012): 1–6, doi:10.1038/srep00540.

77 T. C. Tan et al., "Lifestyle Risk Factors Associated with Threatened Miscarriage: A Case-Control Study," *Journal of Fertilization* 2, no. 123 (2014), doi:http://dx.doi.org/10.4172/jfiv.1000123.

78 Y. Zhang et al., "Effects of Fetal Microwave Radiation Exposure on Offspring Behavior in Mice," *Journal of Radiation Research* 56, no. 2 (2015): 261–268, doi:10.1093/jrr/rru097.

79 Ozlem Sangun et al., "The Effects of Long-Term Exposure to a 2450 MHz Electromagnetic Field on Growth and Pubertal Development in Female Wistar Rats," *Electromagnetic Biology and*

Medicine 34, no. 1 (2015): 63–71, doi:http://dx.doi.org/10.3109/15368378.2013.871619.

80 Fatemeh Shamsi Mahmoudabadi et al., "Use of Mobile Phone During Pregnancy and the Risk of Spontaneous Abortion," *Journal of Environmental Health Science and Engineering* 13 (2015): 34, doi:10.1186/s40201-015-0193-z; Li-ying Zhou et al., "Epidemiological Investigation of Risk Factors of the Pregnant Women with Early Spontaneous Abortion in Beijing," *Chinese Journal of Integrative Medicine* (2015): 1–5, doi:10.1007/s11655-015-2144-z.

81 Ashok Agarwal et al., "Effects of Radiofrequency Electromagnetic Waves (RF-EMW) from Cellular Phones on Human Ejaculated Semen: An In Vitro Pilot Study," *Fertility and Sterility* 92, no. 4 (2009): 1318–1325, doi:http://dx.doi.org/10.1016/j.fertnstert.2008.08.022.

82 Conrado Avendaño et al., "Use of Laptop Computers Connected to Internet through Wi-Fi Decreases Human Sperm Motility and Increases Sperm DNA Fragmentation," *Fertility and Sterility* 97, no. 1 (2012): 39-45, doi:http://dx.doi.org/10.1016/j.fertnstert.2011.10.012.

83 Jessica A. Adams, "Effect of Mobile Telephones on Sperm Quality: A Systematic Review and Meta-Analysis," *Environment International* 70 (2014): 106–112, doi:http://dx.doi.org/10.1016/j.envint.2014.04.015.

84 Saeed Shokri et al., "Effects of Wi-Fi (2.45 GHz) Exposure on Apoptosis, Sperm Parameters and Testicular Histomorphometry in Rats: A Time Course Study," *Cell Journal* 17, no 2 (2015): 322–331, http://www.ncbi.nlm.nih.gov/pmc/articles/PMC4503846/.

85 Maneesh Mailankot et al., "Radio Frequency Electromagnetic Radiation (RF-EMF) from GSM (0.9/1.8GHZ) Mobile Phones Induces Oxidative Stress and Reduces Sperm Motility in Rats," *Clinics* 64, no. 6 (2009): 561–565, doi:10.1590/S1807-59322009000600011.

86 Suleyman Dasdag et al., "Effect of Long-Term Exposure of 2.4 GHz Radiofrequency Radiation Emitted from Wi-Fi Equipment on Testes Functions," *Electromagnetic Biology and Medicine* 34, no. 1 (2015): 37–42, doi:http://dx.doi.org/10.3109/15368378.2013.869752.

87 E. Odaci and C. Özyilmaz, "Exposure to a 900 MHz Electromagnetic Field for 1 Hour a Day over 30 Days Does Change the Histopathology and Biochemistry of the Rat Testis," *International Journal of Radiation Biology* 91, no. 7 (2015): 547–554, doi:10.3109/09553002.2015.1031850.

88 Aminollah Bahaodini et al., "Low Frequency Electromagnetic Fields Long-Term Exposure Effects on Testicular Histology, Sperm Quality and Testosterone Levels of Male Rats," *Asian Pacific Journal of Reproduction* 4, no. 3 (2015): 195–200, doi:http://dx.doi.org/10.1016/j.apjr.2015.06.001.

89 L. Hillert et al., "The Effects of 884MHz GSM Wireless Communication Signals on Headache and Other Symptoms: An Experimental Provocation Study," *Bioelectromagnetics* 29. no. 3 (2008): 185–196,
doi:10.1002/bem.20379.

90 M. N. Halgamuge, "Pineal Melatonin Level Disruption in Humans Due to Electromagnetic Fields and ICNIRP Limits," *Radiation Protection Dosimetry* 154, no. 4 (2013): 405–416, doi:10.1093/rpd/ncs255.

91 Haitham S. Mohammed et al., "Non-Thermal Continuous and Modulated Electromagnetic Radiation Fields Effects on Sleep EEG of Rats," *Journal of Advanced Research* 4, no. 2 (2013): 181–187, doi:http://dx.doi.org/10.1016/j.jare.2012.05.005.

92 Honglong Cao et al., "Circadian Rhythmicity of Antioxidant Markers in Rats Exposed to 1.8 GHz Radiofrequency Fields,"

International Journal of Environmental Research and Public Health 12, no. 2 (2015): 2071–2087, doi:10.3390/ijerph120202071.

93 Y. P. Luo et al., "Effect of American Ginseng Capsule on the Liver Oxidative Injury and the Nrf2 Protein Expression in Rats Exposed by Electromagnetic Radiation of Frequency of Cell Phone," *Chinese Journal of Integrated Traditional and Western Medicine* 34, no. 5 (2014): 575–580, http://www.ncbi.nlm.nih.gov/pubmed/24941847.

94 Luisa Nascimento Medeiros and Tanit Ganz Sanchez, "Tinnitus and Cell Phones: The Role of Electromagnetic Radiofrequency Radiation," *Brazilian Journal of Otorhinolaryngology* 82, no. 1 (2016): 97–104, doi:http://dx.doi.org/10.1016/j.bjorl.2015.04.013.

95 K. Miller et al., "Erythema Ab Igne.," *Dermatology Online Journal* 17, no. 10 (2011): 28, http://www.ncbi.nlm.nih.gov/pubmed/22031654; Sudhir U. K. Nayak, Shrutakirthi D. Shenoi, and Smitha Prabhu, "Laptop Induced Erythema Ab Igne," Indian Journal of Dermatology 57, no. 2 (2012): 131–132, doi:10.4103/0019-5154.94284; R. R. Riahi and P. R. Cohen, "Laptop-Induced Erythema Ab Igne: Report and Review of Literature," Dermatology Online Journal 18, no. 6 (2012): 5, http://www.ncbi.nlm.nih.gov/pubmed/22747929; G. Sharma, "Burn Injury Caused by Laptop Computers," Annals of Medical and Health Sciences Research 3, no. S1 (2013): S31–S32, doi:10.4103/2141-9248.121216.

96 Harold I. Zeliger, "Electromagnetic Radiation and Toxic Exposure," in *Human Toxicology of Chemical Mixtures*, 3rd ed. (2011): 205–217, doi:http://dx.doi.org/10.1016/B978-1-4377-3463-8.00016-3.

Chapter 5

97 "2015 ASHA Children Technology Use Ages 0-8 Survey," *Crux Research, American-Speech-Language-Hearing Association, Honeoye Falls*, 2015, http://www.asha.org/uploadedFiles/BHSM-Parent-Poll.pdf.

98 Independent Expert Group on Mobile Phones, "A Precautionary Approach," in *Mobile Phones and Health (The Stewart Report)* (May 2000): 107–125, http://webarchive.nationalarchives.gov.uk/20101011032547/http://www.iegmp.org.uk/documents/iegmp_6.pdf.

99 Devra Lee Davis et al., "Swedish Review Strengthens Grounds for Concluding That Radiation from Cellular and Cordless Phones is a Probable Human Carcinogen," *Pathophysiology* 20, no. 2 (2013): 123–129, doi:http://dx.doi.org/10.1016/j.pathophys.2013.03.001.

100 S. Thomas et al., "Use of Mobile Phones and Changes in Cognitive Function in Adolescents," *Occupational and Environmental Medicine* 67, no. 12 (2010): 861–866, doi:10.1136/oem.2009.054080.

101 S. Thomas et al., "Exposure to Radio-Frequency Electromagnetic Fields and Behavioural Problems in Bavarian Children and Adolescents," *European Journal of Epidemiology* 25, no. 2 (2010): 135–141, doi:10.1007/s10654-009-9408-x.

102 L. Lloyd Morgan, Santosh Kesari, and Devra Lee Davis, "Why Children Absorb More Microwave Radiation Than Adults: The Consequences," *Journal of Microscopy and Ultrastructure* 2, no. 4 (2014): 197–204, doi:http://dx.doi.org/10.1016/j.jmau.2014.06.005.

103 Meredith Engel, "Hold the Phone, Central! Cell Phone Radiation Can Cause Cancer: Study," *Daily News*, July 29, 2015, http://www.nydailynews.com/life-style/health/cell phone-radiation-cancer-study- article-1.2308509.

104 "Cell Phones, Wireless Devices Connected to Cancer—Study," RT.com, July 31, 2015, https://www.rt.com/usa/311303-cell phone-mobile-cancer-study/.

105 Om P. Gandhi, Gianluca Lazzi, and Cynthia M. Furse, "Electromagnetic Absorption in the Human Head and Neck for Mobile Telephones at 835 and 1900 MHz," *IEEE Transactions on Microwave Theory and Techniques* 44, no. 10 (1996): 1884–1897, http://www.ece.ncsu.edu/erl/html2/papers/lazzi/1996/NCSU-ERL-LAZZI-96-03.pdf.

106 Morrison v. Portland Public Schools: United States District Court, District of Oregon, Portland Division, Civil Action No. 3:11-cv-00739-MO (2012), http://apps.fcc.gov/ecfs/document/view?id=7520958125.

107 Scott O'Connell, "Family Sues Fay School in Southboro, Claims Wi-Fi Made Son Ill," *Telegram and Gazette*, August 24, 2015, http://www.telegram.com/article/20150824/NEWS/150829606.

108 Letter to Fay School Board of Trustees, District of Massachusetts, Worcester Division, Case 4:15-cv-40116-TSH (2015), http://www.casewatch.org/civil/wifi/complaint.pdf.

109 "iPad User Manual's Safety Warning and Disclaimer. Have You Read It?" *Is Wi-Fi Safe for Children? Beware of Health Risks* (blog), April 7, 2013, http://www.safeinschool.org/2013/04/ipad-user-manuals-safety-warning-and.html.

110 Jenny S. Radesky, Jayna Schumacher, and Barry Zuckerman, "Mobile and Interactive Media Use by Young Children: The Good, the Bad, and the Unknown," *Pediatrics* 135, no. 1 (2015), doi:10.1542/peds.2014-2251.

111 "Media and Children Toolkit," American Academy of Pediatrics, https://www.aap.org/en-us/advocacy-and-policy/aap-health-initiatives/Pages/Media-and-Children.aspx.

112 Y. Zhang et al., "Effects of Fetal Microwave Radiation Exposure on Offspring Behavior in Mice," Journal of Radiation Research 56, no. 2 (2015): 261–268, doi:10.1093/jrr/rru097.

113 Jean-Pierre Marc-Vergnes, "Electromagnetic Hypersensitivity: The Opinion of an Observer Neurologist," *Comptes Rendus Physique* 11, no. 9 (2010): 564–575, doi:10.1016/j.crhy.2010.12.006.

114 Stephen J. Genuis and Christopher T. Lipp, "Electromagnetic Hypersensitivity: Fact or Fiction?" *Science of the Total Environment* 414 (2012): 103–112, doi:http://dx.doi.org/10.1016/j.scitotenv.2011.11.008.

115 Olle Johansson, *Evidence for Effects on the Immune System* (July 2007), http://www.bioinitiative.org/report/wp-content/uploads/pdfs/sec08_2007_Evidence_%20Effects_%20Immune_System.pdf.

116 M. Röösli et al., "Symptoms of Ill Health Ascribed to Electromagnetic Field Exposure—A Questionnaire Survey," *International Journal of Hygiene and Environmental Health* 207, no 2 (2004): 141–150, doi:10.1078/1438-4639-00269.

117 Susan Parsons, "Living with Electrohypersensitivity: A Survival Guide," WEEP Initiative, http://www.weepinitiative.org/livingwithEHS.html.

118 Joel Moskowitz, "Electromagnetic Hypersensitivity," *Electromagnetic Radiation Safety* (blog), September 1, 2016, http://www.saferemr.com/2014/10/electromagnetic-hypersensitivity_30.html.

119 Susan Parsons, "Living with Electrohypersensitivity: A Survival Guide," WEEP Initiative, http://www.weepinitiative.org/livingwithEHS.html.

120 "Gadget 'Allergy': French Woman Wins Disability Grant," BBC News, August 27, 2015, http://www.bbc.com/news/technology-34075146.

121 Y. Zhang et al., "Effects of Fetal Microwave Radiation Exposure on Offspring Behavior in Mice," Journal of Radiation Research 56, no. 2 (2015): 261–268, doi:10.1093/jrr/rru097.

122 Helke Ferrie, "Creating a Healthy Home: Strategies for Removing or Reducing Dirty Electricity and EMF Radiation," *Vitality* magazine, http://vitalitymagazine.com/article/creating-a-healthy-home/.

Chapter 6

123 Steven W. Smith, "The Breadth and Depth of DSP," in *The Scientist and Engineer's Guide to Digital Signal Processing* (San Diego, CA: California Technical Publishing, 1997), http://www.dspguide.com/ch1/1.htm.

124 "Milestones: Speak and Spell, the First Use of a Digital Signal Processing IC for Speech Generation, 1978," Engineering and Technology History Wiki, last modified December 30, 2015, http://ethw.org/Milestones:Speak_&_Spell,_the_First_Use_of_a_Digital_Signal_Processing_IC_for_Speech_Generation,_1978.

125 Baby In The Corner, "Speak and Spell," *Films of the 80s* (blog), June 7, 2014, http://social.rollins.edu/wpsites/filmsofthe80s/2014/06/07/speak-spell/.

126 "Specific Absorption Rate (SAR) for Cellular Telephones," FCC, last modified November 30, 2015, https://www.fcc.gov/encyclopedia/specific-absorption-rate-sar-cellular-telephones.

127 "What is SAR and What is all The Fuss About?" SAR Values, http://sarvalues.com/what-is-sar-and-what-is-all-the-fuss-about/.

128 Ibid.

129 "SAR and Mobile Phone Safety," AMTA, http://www.amta.org.au/sar.

130 Kenneth Chang, "Debate Continues on Hazards of Electromagnetic Waves," New York Times, July 7, 2014, http://www.nytimes.com/2014/07/08/science/debate-continues-on-hazards-of-electromagnetic-waves.html.

131 Varshini Karthik et al., "Study of the Thermal Effects of EM Radiation from Mobile Phones on Human Head Using IR Thermal Camera," *ITSI Transactions on Electrical and Electronics Engineering* 3, no. 1 (2015): 24–26, http://www.irdindia.in/journal_itsi/pdf/vol3_iss1/5.pdf.

132 Ibid.

133 Ronald N. Kostoff and Clifford G. Y. Lau, "Combined Biological and Health Effects of Electromagnetic Fields and Other Agents in the Published Literature," *Technological Forecasting and Social Change* 80, no. 7 (2013): 1331–1349, doi:http://dx.doi.org/10.1016/j.techfore.2012.12.006.

134 Priya Ganapati, "Inside a Cellphone Radiation Testing Lab," Wired, October 27, 2009, https://www.wired.com/2009/10/cellphone-radiation-testing/.

135 Om P. Gandhi et al., "Exposure Limits: The Underestimation of Absorbed Cell Phone Radiation, Especially in Children," *Electromagnetic Biology and Medicine* 31, no. 1 (2012): 34–51, doi:http://dx.doi.org/10.3109/15368378.2011.622827.

136 "FCC's Cell Phone Testing Dummy is Larger Than 97% of All Cell Phone Users," Consumers for Safe Phones, November 27, 2011, http://consumers4safephones.com/

fccs-cell-phone-testing-dummy-is-larger-than-97-of-all-cell-phone-users/.

137 "Bioinitiative 2012 Media," BioInitiative Working Group, http://www.bioinitiative.org/media/.

138 "Specific Absorption Rate (SAR) For Cell Phones: What It Means for You," FCC, last modified October 25, 2016, https://www.fcc.gov/guides/specific-absorption-rate-sar-cell-phones-what-it-means-you.

139 "Bioinitiative 2012 Media," BioInitiative Working Group, http://www.bioinitiative.org/media/.

140 Cindy Sage and David O. Carpenter, eds., *BioInitiative Report 2012* (n.p.: BioInitiative Working Group, 2012), http://www.bioinitiative.org.

141 "FCC Takes Steps to Facilitate Mobile Broadband and Next Generation Wireless Technologies in Spectrum above 24 GHz," FCC, July 14, 2016, https://apps.fcc.gov/edocs_public/attachmatch/DOC-340301A1.pdf.

142 Vic Micolucci, "New FCC Rule Allows for Next Generation Wireless Technology," News4Jax, July 15, 2016, http://www.news4jax.com/tech/new-fcc-rule-allows-for-next-generation-wireless-tech.

143 Josh Del Sol, "FCC Intimidates Press and Evades Questioning About Wireless and Cancer at 5G Rollout," Collective Evolution, July 28, 2016, http://www.collective-evolution.com/2016/07/28/gestapo-in-the-usa-5g-fcc-intimidates-press-and-kills-free-speech/.

144 Ibid.

145 Tom Wheeler, "National Press Club Luncheon with FCC Chair Tom Wheeler" (press conference, National Press Club, Washington,

DC, June 20, 2016), http://www.press.org/sites/default/files/20160620_wheeler.pdf.

146 Ibid.

Chapter 7

147 Jean Elle, "Berkeley Approves 'Right to Know' Cell Phone Radiation Warning Ordinance," NBC Bay Area—KNTV, May 12, 2015, http://www.nbcbayarea.com/news/local/Berkeley-Approves-Right-to-Know-Cell-Phone-Ordinance-303551751.html.

148 Cyrus Farivar, "City of Berkeley Fends Off Wireless Industry Suit Over Health Warning," Ars Technica, September 21, 2015, http://arstechnica.com/tech-policy/2015/09/city-of-berkeley-fends-off-wireless-industry-suit-over-health-warning/.

149 Cyrus Farivar, "Berkeley's Cell Phone Radiation Warning Law Can Go into Effect, Judge Rules," Ars Technica, January 28, 2016, http://arstechnica.com/tech-policy/2016/01/berkeleys-cell-phone-radiation-warning-law-can-go-into-effect-judge-rules/.

150 Jean Elle, "Berkeley Approves 'Right to Know' Cell Phone Radiation Warning Ordinance," NBC Bay Area—KNTV, May 12, 2015, http://www.nbcbayarea.com/news/local/Berkeley-Approves-Right-to-Know-Cell-Phone-Ordinance-303551751.html.

151 George Carlo and Martin Schram, "Cell Phones: Invisible Hazards in the Wireless Age" (New York: Carroll & Graf, 2001), https://books.google.com/books?id=LwtNzv14KdgC&printsec=frontcover#v=onepage&q&f=false.

152 Robert N. Proctor, "Tobacco and Health," *Journal of Philosophy, Science and Law* 4 (2004), http://jpsl.org/archives/

tobacco-and-health-expert-witness-report-filed-behalf-plaintiffs-united-states-america-plaintiff-v-philip-morris-inc-et-al-defen/.

153 "Smoking and Tobacco Use," Centers for Disease Control and Prevention, last modified November 15, 2012, http://www.cdc.gov/tobacco/data_statistics/by_topic/policy/legislation/.

154 "iPhone 7 RF Exposure information," Apple, http://www.apple.com/legal/rfexposure/iphone9,1/en/.

Chapter 8

155 "Welcome to the Precautionary Principle Website," The Precautionary Principle, http://www.precautionaryprinciple.eu/.

156 "Radiation Dispersal from Japan," CDC, last modified March 21, 2011, http://www.cdc.gov/niosh/topics/radiation/radbasics.html.

Chapter 9

157 "Nancy Wertheimer, Who Linked Magnetic Fields to Childhood Leukemia, Dies," Microwave News, January 23, 2008, http://microwavenews.com/news-center/nancy-wertheimer-who-linked-magnetic-fields-childhood-leukemia-dies.

158 Yue Wang, "More People Have Cell Phones Than Toilets, U.N. Study Shows," *Time*, March 25, 2013, http://newsfeed.time.com/2013/03/25/more-people-have-cell-phones-than-toilets-u-n-study-shows/

159 Lee Rainie, "Cell Phone Ownership Hits 91% of Adults," Pew Research Center, June 6, 2013, http://www.pewresearch.org/fact-tank/2013/06/06/cell-phone-ownership-hits-91-of-adults/.

160 "Device Ownership Over Time," Pew Research Center, http://www.pewinternet.org/data-trend/mobile/device-ownership/.

161 I. Yakymenko et al., "Oxidative Mechanisms of Biological Activity of Low-Intensity Radio Frequency Radiation," *Electromagnetic Biology and Medicine* 35, no. 2 (2016): 196–202, doi:10.3109/15368378.2015.1043557.

162 Mary Bellis, "History of Laptop Computers," About.com Inventors, last modified August 6, 2016, http://inventors.about.com/library/inventors/bllaptop.htm.

163 "Device Ownership Over Time," Pew Research Center, http://www.pewinternet.org/data-trend/mobile/device-ownership/.

164 Sascha De Gersdorff, "Health Risks of Mobile Devices," *Women's Health*, October 1, 2010, http://www.womenshealthmag.com/sex-and-love/cell-phone-radiation.

165 Geoffry N. De Iuliis et al., "Mobile Phone Radiation Induces Reactive Oxygen Species Production and DNA Damage in Human Spermatozoa *In Vitro*," *PLOS ONE* 4, no. 7 (2009), doi:http://dx.doi.org/10.1371/journal.pone.0006446.

166 L. Hillert et al., "The Effects of 884MHz GSM Wireless Communication Signals on Headache and Other Symptoms: An Experimental Provocation Study," Bioelectromagnetics 29. no. 3 (2008): 185–196, doi:10.1002/bem.20379.

167 Julie Bort, "The History of the Tablet, an Idea Steve Jobs Stole and Turned Into a Game-Changer," *Business Insider*, June 2, 2013, http://www.businessinsider.com/history-of-the-tablet-2013-5?op=1.

168 "Shipment Forecast of Laptops, Desktop PCs and Tablets Worldwide from 2010 to 2019 (in Million Units),"

Statista, 2016, https://www.statista.com/statistics/272595/global-shipments-forecast-for-tablets-laptops-and-desktop-pcs/.

169 "Device Ownership Over Time," Pew Research Center, http://www.pewinternet.org/data-trend/mobile/device-ownership/.

170 "Kids Wireless Use Facts," Growing Wireless, http://www.growingwireless.com/get-the-facts/quick-facts.

171 US Department of Energy, *2014 Smart Grid System Report* (Washington, DC: US Department of Energy, 2014), http://energy.gov/sites/prod/files/2014/08/f18/SmartGrid-SystemReport2014.pdf.

172 "How Many Smart Meters Are Installed in The United States, and Who Has Them?," US Energy Information Administration, last modified April 25, 2016, http://www.eia.gov/tools/faqs/faq.cfm?id=108&t=3.

173 "Smart Meters," EMF Safety Network, http://emfsafetynetwork.org/smart-meters/.

174 Ibid.

175 F. Lamech, "Self-Reporting of Symptom Development from Exposure to Radio Frequency Fields of Wireless Smart Meters in Victoria, Australia: A Case Series," *Alternative Therapies in Health and Medicine* 20, no. 6 (2014): 28–39, http://www.ncbi.nlm.nih.gov/pubmed/25478801.

176 "Wireless Utility Meter Safety Impacts Survey," *SurveyDNA.com, EMF Safety Network*, 2011, http://emfsafetynetwork.org/wp-content/uploads/2011/09/Wireless-Utility-Meter-Safety-Impacts-Survey-Results-Final.pdf.

177 Ibid.

178 David Watkins, "Global Broadband and WLAN (Wi-Fi) Networked Households Forecast 2010–2019," Strategy Analytics, December 18, 2015, https://www.strategyanalytics.com/access-services/devices/connected-home/consumer-electronics/reports/report-detail/global-broadband-and-wlan-(wi-fi)-networked-households-forecast-2010-2019.

179 Letter to Fay School Board of Trustees, District of Massachusetts, Worcester Division, Case 4:15-cv-40116-TSH (2015), http://www.casewatch.org/civil/wifi/complaint.pdf.